The Experimental City

'If the much talked about politics of green transformations during the post-2009 era is to be more than another fad, it will have to find traction in the world's cities where the majority of the population now lives. This book convincingly argues that this is starting to happen as "urban experiments" mushroom across the global regions of the world. This first provocative review of the evolutionary potential of these actually existing processes of change starts off in Chapter 2 with a remarkably useful definition of "urban experimentation" that is then explored by people from many different disciplines and perspectives. The analyses emerging from this book raise the possibility that "urban experimentation" may well be the emergent mode of governance that replaces both bureaucratic managerialism and the business-led public–private partnerships that underpinned neoliberal corporatism and splintered urbanism. Entrepreneurs, innovators and knowledge networks become the new players in a world where cities become laboratories for the future.'

Mark Swilling, *Distinguished Professor of Sustainable Development in the School of Public Leadership, Stellenbosch University, South Africa*

'From Peñalosa to post carbon cities, from urban acupuncture to Arcosanti and from living labs to cabin ecologies, experimentation is interrogated in this creative and innovative collection through multiple cases focusing on governance, conceptual development, stakeholder engagement, and corporatist approaches. At the intersection of practice and theory, the contributors bring us closer to identifying common concerns, challenges, and questions.'

Julian Agyeman, *Professor of Urban and Environmental Policy and Planning, Tufts University, USA*

This book explores how the concept of urban experimentation is being used to reshape practices of knowledge production in urban debates about resilience, climate change governance, and socio-technical transitions.

With contributions from leading scholars, and case studies from the Global North and South, from small- to large-scale cities, this book suggests that urban experiments offer novel modes of engagement, governance, and politics that both challenge and complement conventional strategies. The book is organized around three cross-cutting themes. Part I explores the logics of urban experimentation, different approaches, and how and why they are deployed. Part II considers how experiments are being staged within cities, by whom, and with what effects. Part III examines how entire cities or groups of cities are constructed as experiments.

This book develops a deeper and more socially and politically nuanced understanding of how urban experiments shape cities and drive wider changes in society, providing a framework to examine the phenomenon of urban experimentation in conceptual and empirical detail.

James Evans is a Professor of Geography at the School of Environment, Education and Development, University of Manchester, UK.

Andrew Karvonen is Assistant Professor in Urban and Regional Studies at the KTH Royal Institute of Technology, Sweden.

Rob Raven is Professor of Institutions and Transitions at Utrecht University, the Netherlands.

Routledge research in sustainable urbanism

This series offers a forum for original and innovative research that engages with key debates and concepts in the field. Titles within the series range from empirical investigations to theoretical engagements, offering international perspectives and multidisciplinary dialogues across the social sciences.

Co-producing Knowledge for Sustainable Cities
Joining forces for change
Edited by Merritt Polk

Global Garbage
Urban imaginaries of waste, excess, and abandonment
Edited by Christoph Lindner and Miriam Meissner

The Experimental City
Edited by James Evans, Andrew Karvonen and Rob Raven

The Experimental City

Edited by
James Evans, Andrew Karvonen
and Rob Raven

Routledge
Taylor & Francis Group
LONDON AND NEW YORK

First published in paperback 2018

First published 2016
by Routledge
2 Park Square, Milton Park, Abingdon, Oxon OX14 4RN

and by Routledge
711 Third Avenue, New York, NY 10017

Routledge is an imprint of the Taylor & Francis Group, an informa business

British Library Cataloguing-in-Publication Data
A catalogue record for this book is available from the British Library

Library of Congress Cataloging-in-Publication Data
Names: Evans, James Philip Martin, editor. | Karvonen, Andrew, 1971– editor. | Raven, Rob, 1975– editor.
Title: The experimental city / edited by James Evans, Andrew Karvonen and Rob Raven.
Description: Abingdon, Oxon ; New York, NY : Routledge, 2016. |
Series: Routledge research in sustainable urbanism
Identifiers: LCCN 2015047070| ISBN 9781138856202 (hardback) |
ISBN 9781315719825 (e-book)
Subjects: LCSH: Sustainable urban development. | Community development, Urban. | Urban policy.
Classification: LCC HT241 .E966 2016 | DDC 307.76—dc23
LC record available at http://lccn.loc.gov/2015047070

ISBN: 978-1-138-85620-2 (hbk)
ISBN: 978-1-138-29967-2 (pbk)
ISBN: 978-1-315-71982-5 (ebk)

Typeset in Times New Roman
by Book Now Ltd, London

Contents

Figures

Tables

Notes on contributors

Frans Berkhout is Executive Dean of the Faculty of Social Science and Public Policy, and Professor of Environment, Society and Climate in the Department of Geography at King's College London. From 2013 to early 2015, he served as director of the Future Earth programme, based at the International Council for Science (ICSU) in Paris. His recent work has been concerned with science, technology, policy and sustainability, with a focus on climate change.

Lynsay Blake is a Research Coordinator in the Institute for Sustainability at Newcastle University. Her research interests include the development of environmentally and economical sustainable solid-waste treatment systems, and collaborative research across social and technical disciplines.

Vanesa Castán Broto is a senior lecturer in UCL's Bartlett Development Planning Unit with an interdisciplinary background in engineering and environmental sociology. Her research focuses on how technology and environmental knowledge mediate the relationship between society and the environment and the impact of these processes in urban and regional planning.

Federico Cugurullo is a lecturer in Human Geography and Sustainable Urbanism at the University of Manchester. His research is positioned at the intersection of urban geography and political philosophy, and explores how ideas of sustainability are cultivated and implemented across geographical spaces, with a focus on projects for eco-cities.

James Evans is a Professor of Geography at the School of Environment, Education and Development, University of Manchester. His core research interest is in urban environmental governance with a substantive focus on how cities learn to become more sustainable in the face of multiple challenges. His current research projects focus on urban living labs, smart cities and resilience.

Idil Gaziulusoy is Principal Researcher at Victorian Eco-Innovation Lab at the University of Melbourne. She has a background in engineering, industrial design and sustainability science. Her broad research interest is in design and innovation for sustainability specifically contributing to the development of theory and practice in the emerging field of design for system innovations and transitions.

Bipashyee Ghosh is a doctoral researcher in the Science Policy Research Unit (SPRU) at the University of Sussex. Her research interest lies in urban mobility transition in India, critically reflecting on issues such as environmental sustainability and social inclusion.

Duke Ghosh is a researcher at Global Change Research in Kolkata. His current research interests are sustainability transitions and policy analysis.

Maarten Hajer is Professor of Urban Futures at Utrecht University and the Chief Curator of the 2016 International Architecture Biennale Rotterdam (IABR). He is co-chairing the working group on Cities of UNEP's International Resource Panel.

Mike Hodson is based jointly in the Sustainable Consumption Institute and the Manchester Institute of Innovation Research at the University of Manchester. His developing research interests are at the interface of systemic transitions and territorial transitions. In particular Mike's research focuses on the relationships between systemic transitions, territorial transitions and the ways in which relationships between the two are, are not and can be organized.

Andrew Karvonen is Assistant Professor in Urban and Regional Studies at the KTH Royal Institute of Technology, Sweden and co-director of the Centre for Urban Resilience and Energy. He conducts research on the political and cultural aspects of urban infrastructure with a particular emphasis on sustainable transitions.

Katherine Luke studied environmental governance at the University of Manchester. Her research focuses on environmental history, urban regeneration, public health, and alternative models of local economic development in post-industrial regions of the Global North. She briefly worked in agriculture at Arcosanti and has continued to be involved in developing community-managed green spaces.

Kes McCormick is an Associate Professor and Assistant Head at the International Institute for Industrial Environmental Economics (IIIEE) at Lund University in Sweden. With a background in political science and environmental science, he engages in a combination of research, education and innovation activities in the fields of sustainability and governance. He focuses on how to accelerate the implementation of renewable energy technologies and energy efficiency improvements, catalyze sustainable urban transformation, and promote education for sustainability.

Simon Marvin is Professor and Director of the Urban Institute at the University of Sheffield. His research interests focus on the changing relationships between cities and infrastructure networks. More recently his work has focused on comparative analysis of low carbon transitions and the interactions between smart technologies and urban contexts.

Tim May is Professor in the Centre for Sustainable Urban and Regional Futures, University of Salford, Manchester. Tim's intellectual interests focus on knowledge generation and reception, sustainability and urban change. Tim has authored 14 books and over 180 articles, book chapters and other publications. With Beth Perry, he is currently writing three books on 'Reflexivity', 'Cities and Knowledge' and 'Social Research'. He is also writing a third edition, with Zygmunt Bauman, of *Thinking Sociologically* and articles on class, governance, urban crisis and knowledge.

Beth Perry is Reader and Director of UPRISE/Sustainable Urban and Regional Futures, in the School of the Built Environment, University of Salford. Her research focuses on critically interrogating and developing pathways to more just sustainable urban futures, with an emphasis on urban governance and the roles of universities and academics. She has a sustained record of research performance at national and international levels and is currently writing three books with Tim May (*Cities and the Knowledge Economy*, Earthscan/Routledge; *Reflexivity: A Guide*, Sage; and *Social Research: Issues, Methods and Process*, *5th Edition*, McGraw Hill/Open University Press.

Stephanie Pincetl is a Research Professor at the Institute of the Environment and Sustainability at UCLA. She directs the California Center for Sustainable Communities (CCSC) and does research on energy and water flows, ecosystem services, and ideas of nature. Her Center does science for the public interest and is highly interdisciplinary.

Gareth Powells is a lecturer in Human Geography at Newcastle University. His research and teaching focus on the political ecologies of urban sustainability and socio-technical studies of transition and innovation processes.

Elizabeth Rapoport is an interdisciplinary urban researcher whose work focuses on the intersection of sustainability and climate change objectives with urban planning, policy and governance. Currently, Elizabeth is a Research Associate with the City Leadership Initiative in the Department of Science, Technology, Engineering and Public Policy at University College London. Elizabeth holds an EngD from UCL and an MSc from the London School of Economics, and has worked as a practitioner on urban planning and policy in numerous countries around the world.

Rob Raven is Professor of Institutions and Societal Transitions at Utrecht University.

Joyashree Roy is an ICSSR national fellow and Professor of Economics as well as coordinator of the Global Change Programme and Director of the JU-Sylff Programme of Jadavpur University in Kolkata, India. Her current research interests include economics of low carbon growth trajectories, coastal ecosystem services, and challenges in social embedding of technology in developing countries.

Chris Ryan is Director of the Victorian Econ-Innovation Lab and Professor of Urban Eco-Innovation at the University of Melbourne. He is the chief investigator for the national Australian Visions and Pathways 2040 research project and a member of the leadership team of the Future City research program in the University of Melbourne Sustainable Society Institute. He is a visiting Professor at TUDelft in the Netherlands and a past Director of the International Institute for Industrial Environmental Economics at Lund University in Sweden and was founding Director of the national Centre for Environmental Design at RMIT University in Melbourne.

Martin Sanzana Calvet is a doctoral candidate in the Development Planning Unit at University College London. His research is focused on the urban political ecology of cities under change and conflict.

Gabriele Schliwa is a PhD researcher at the School for Environment, Education and Development at the University of Manchester. Her research focusses on public participation in the context of smart city governance, living labs and sustainable urban mobility.

Frans Sengers is a postdoctoral researcher at Utrecht University. His current research about transformative socio-technical change focuses on smart city and eco-city developments in Europe and China.

Sofia T. Shwayri is a Visiting Scholar at the University of California at Berkeley. Formerly, she was an Associate Professor at Seoul National University in South Korea. Sofia's research is focused on understanding Korea's sustainable practices in cities as diverse as the capital Seoul, the administrative city of Sejong and the planned eco-city of Songdo. Her studies also include the role of war and post-war reconstruction in the making of the contemporary Middle Eastern city.

Jonathan Silver is a Leverhulme Early Career Fellow at the Department of Geography, Durham University, working on the politics of urban infrastructure.

Michael Trudgeon is the Deputy Director of the Victorian Eco Innovation Lab at the University of Melbourne. He is responsible for delivering the Eco-acupuncture design studio teaching program to Masters of Architecture program at UoM. He is design director at Crowd Productions, a Melbourne-based architecture and industrial design studio. His practice and research background is in developing strategies to prototype new technology and spatial solutions for commercial architecture projects.

Jana Wendler is a postdoctoral researcher in Human Geography at the University of Manchester. Her work focuses on creative, experimental and playful approaches to urban change and geographical learning.

Anna J. Wieczorek is a lecturer in the School of Innovation Sciences at the Eindhoven University of Technology. She researches how sustainability transitions are motivated and conceived in various economic and geographical contexts with a particular focus on emerging economies. She is also involved in the translation of transitions theories to practice and academic as well as professional education.

Thomas Yarrow is a senior lecturer in Social Anthropology at Durham University. His work focuses on expert knowledge as social practice, particularly in relation to built space.

Foreword

Maarten Hajer

Cities are increasingly seen as agents of change. Benjamin Barber authored *If Mayors Ruled the World* and travels the globe promoting a 'Parliament of Mayors'. Others speak of an 'urban age' and try to connect cities through networks. Is this idea of an urban age more than just a nice story line? We see how national governments struggle to come up with policy solutions for the pressing problems of our times, such as the climate crisis, migration, the future of work and the future of health and care. Nation states are an invention of the nineteenth century. Would it be that cities are more apt to meet the challenges of the twenty-first century? In actual fact there is more going on than this profound reordering of the political geography.

Cities had no option but to reinvent themselves over the last decades. First there was the normative call for a different form of government. The 1990s saw the emergence of the New Public Management approach with the dictum 'run government like a business'. This paradigm caused, and in many cases still causes, massive confusion. Perhaps it was good to shake up the old bureaucracy and insert a dose of ideas from the world of business management. But a public and democratic organization is something essentially different from an organization operating in the world of business and commerce.

Second, in many countries local governments got more responsibilities, for instance in the field of social policy, health care and welfare, as national governments tried to reduce its own involvement in those areas.

Third, local governments have been through long years of budget cuts that have made it impossible to keep relevant expertise in house. City governments no longer know how to judge the technical proposal for a new bridge, a town hall, a new ICT system. All expertise has been outsourced to consultants. As a consequence the bureaucratic capacity of the state to rule has been eroded. So, while local governments now often have more responsibilities, they simply cannot draw on the old logics of their bureaucratic routines. Local governments have to do more with less.

No wonder, then, that we hear so much about innovative governance at the local level. Cities simply *must* innovate in order to function. Little wonder that we hear so much about experiments these days. A remarkable example is the rise of 'living labs'. Quite tellingly, the European Network of Living Labs (ENOLL) alone has more than 250 members. It is just one example of a trend of what the editors of this book have called 'experimental cities'. Living labs are part of a whole new lexicon of governance, including other terms and concepts like 'innovation hubs', 'city deals' and 'green deals'. Cities are looking for new trajectories to find solutions for the pressing issues of our times.

This new discourse marks a shift away from old planning practices such as 'survey–analysis–plan' and indeed tools like strategic plans that were to be updated every four or five years. Such practices assume stable bureaucracies that cities, after all the budget cuts, no longer have. But there is another, more profound, reason for this shift. There may simply be less of a need for those cumbersome bureaucratic routines as digitalization now allows much easier access to data, and, indeed, more frequent update of the figures and results. This new digital layer in policy-making disrupts fundamental routines of governance. Indeed, the very basic idea of a 'policy process' has had its day. The differentiation between policy analysis, policy-making, implementation and subsequent evaluation cannot cope with ever quicker feedbacks that stem from ICT applications. Take the example of the national statistical office. In the old days, a national statistical office had to rely on cumbersome questionnaires to gather data. Every bakery was obliged to fill out questionnaires as to how much flour was bought, how much bread was sold, how much energy was used. Adding this up was a horrific task, requiring many 'clerks'. Nowadays this can all be done quite easily and the results are in much quicker. The effect is a push for faster responses and adjustments, thus cancelling out the power of the hierarchically ordered bureaucratic regime. So, like it or not, we must speed up and innovate the way in which city governments respond to and interact with new findings and possibilities.

What is more, there is now a very new and demanding public that is far less patient and calls for quick responses. This broad sociological shift that I have called the rise of 'the energetic society' (2010) also implies that governments have had to readjust their practices of governance to retain their legitimacy.

This book is a timely exercise, taking stock of the many different ways in which these experimental cities have manifested themselves. It provides a much better understanding of the consequence of this turn to experimental governance. A first and important observation is that experimental cities allow for the contextual understanding of issues. There is no 'one size fits all', but working with an appreciation of the local situation. Second, there is no lack of ambition. We see how many experiments are aimed to bring about structural transformation; for instance in the field of sustainable energy.

The idea of 'experimental cities' seems rooted in the 'local turn' that emerged in the aftermath of the 1992 Rio de Janeiro sustainability conference. It launched the 'Local Agenda 21' programme. Initially this was not a prominent stream of initiatives (indeed, it seemed to serve as a fig leaf for the lack of robust actions of national governments). Yet over the years it matured into the now generally accepted claim that cities are new agents of change on a global level.

A third, and perhaps more profound, dimension of these new experimental cities seems to be the shift towards a new epistemology. Rather than trying to apply a general principle at the local level we now hear much more about 'best practices' and 'upscaling'. So a deductive logic seems to have given way to a far more inductive way of reasoning. In terms of learning strategies the old regime of 'analysis & instruction' has been replaced by an approach of 'variation & selection'. In this latter system, there is much more room and appreciation for localized experiments.

At the same time the new discourse of 'living labs' and 'experiments' is a somewhat ironic trend. After all, living in a city that is functioning as a 'living lab' could also be quite a scary proposition. Anthony Townsend (2013) has some examples of how early adoption of smart technologies can also lead to blackouts, accidents or simple disruptions of services.

The explanation for the emergence of this new language may lie in the fact that we have also seen an influx of natural science expertise into urban matters. Traditionally the urban was strongly the domain of the social sciences. But the approaches of human geography, urban studies, urban sociology or urban politics now must compete with new claims to knowledge. This is sometimes via interdisciplinary fields like innovation studies or inquiries into urban metabolism, but we also witness the advance of a new 'urban science' that conceives of the city as an algorithm and tries to extend its claims using the newly available 'big data'.

It seems likely that the natural science disciplines brought this new vocabulary of labs and experiments with them. The risk, of course, is that with the new language you also bring in some of the connected epistemological commitments. There are quite a few remnants of the positivist 'science for policy' approach in the language of experimental cities. Talk of 'samples', 'rigorous scientific knowledge', 'upscaling', 'replicating', 'roll out' all derive from a scientific discourse that assumes that you can 'isolate' crucial variables, come to Humean general laws and then indeed, 'roll out' the model that is supposed to work in any context. Perhaps this 'slippage' is characteristic for this moment in thinking about new theories of change.

Yet in the end behind this linguistic turn of 'experimental cities' is a profound move towards a new way of thinking about social change in which cities are places of hope. This timely book will help us to be more precise about how cities can live up to that great expectation.

References

Hajer, M. (2011). *The Energetic Society: in Search of a Governance Philosophy for a Clean Economy*. The Hague: PBL Netherlands Environmental Assessment Agency.

Townsend, A.M. (2013). *Smart Cities: Big Data, Civic Hackers, and the Quest for a New Utopia*. New York: WW Norton & Company.

1 The experimental city

New modes and prospects of urban transformation

James Evans, Andrew Karvonen and Rob Raven

The world hates change yet it is the only thing that has brought progress.
(Charles Kettering)

The promise of experimentation

As the global population becomes increasingly urbanised, cities have emerged as the dominant arenas to address the grand challenges facing humanity. Problems associated with climate change, economic under-development and social inequality are essentially urban in character. And so are their solutions. The burgeoning realisation that 'business as usual' will no longer do has prompted a search for alternative ways to organise, plan, manage, and live in cities. Experimentation promises a way to do this, gaining traction in cities all over the world as a mode of governance to stimulate alternatives and steer change (Bulkeley and Castán Broto 2012). Policy makers, designers, private companies and third sector organisations are initiating innovation activities to trial alternative future visions of local economic development, social cohesion, environmental protection, creative sector expansion, policy evolution, service delivery, infrastructure provision, academic research, and more (Karvonen *et al.* 2014).

The concept of experimentation feeds on attractive notions of innovation and creativity (both individual and collective) while reframing the emphasis of sustainability from distant targets and government policies to concrete and achievable actions that can be undertaken by a wide variety of urban stakeholders in specific places (Karvonen and van Heur 2014). The ability of urban experiments to be radical in ambition while limited in scope underpins a vibrant debate in both the policy and academic worlds with respect to their ability to prompt genuine change. Are they simply extensions of business as usual, spatially limited and captured by a familiar cast of dominant interests? Or can they generate real alternatives and stimulate profound transformation? The profusion of major international research projects currently addressing this very question suggests that there is no simple answer (see Castán Broto and Bulkeley 2013; McGuirk *et al.* 2014; Wieczorek *et al.* 2015).

The goal of this book is to make sense of urban experimentation as a rapidly emerging field of practice and theory by bringing different approaches and cases into dialogue with one another. To set the scene, this introductory chapter surveys key themes that animate urban experimentation as a field of study before introducing the contributions that comprise the volume.

While an ethos of experimentation has arguably animated the urban project from classical antiquity onwards, urban experiments are currently being deployed more widely and explicitly than at any other time (Evans 2011). The Mayor of Bogatá, Enrique Peñalosa, is famous for transforming the city from one of the most dangerous and unpleasant places to live into a global leader in sustainable urbanism by the end of the 1990s. What he calls the 'Bogatá experiment' started with a far smaller one, the *dia sin carro* or day without cars, which catalysed the global 'Happy City' movement (Montgomery 2013). Real-world experiments can create powerful shared values by letting people experience a different possible future – in this case a city with no cars. This is just one of a multitude of urban movements that share a commitment to changing the way in which we build, manage and live in cities through explicitly staging experiments. Smart cities, eco-cities, low carbon urbanism, urban living labs, happy cities and sustainable urban development all draw on the idea that experimentation can generate more liveable, prosperous and sustainable urban futures (de Jong *et al.* 2015). Experimentation forms a common thread running through otherwise disparate contemporary urban trends, from corporatised attempts to create smart, low carbon cities to grassroots civic movements to make neighbourhoods more socially cohesive. It is for this reason that urban experimentation has come rapidly to prominence across a broad spectrum of urban practice and thought.

While assuming many forms, urban experimentation can be distinguished conceptually from conventional urban development or policy by an explicit emphasis on learning from real-world interventions. Urban experimentation offers a framework within which to arrange instruments, materials and people to induce change in a controlled manner, and subsequently evaluate and learn from those changes (Karvonen and van Heur 2014). The institutionalisation of experimentation sets contemporary activities apart from more broadly experimental approaches to urbanism practiced in previous decades and, indeed, as Gross suggests (2010: 66) 'anything that is subject to change'. This ethos of experimentation resonates with the broader emergence of reflexive governance and the importance of learning within and between networks of urban actors (McFarlane 2011a, 2011b). Current attempts in cities to learn through place-based experimentation reflect Beck's (1995: 15) model of reflexive modernity, seeking to reconcile 'the science of data and the science of experience' through real-world experiments. Research inspired by the laboratory studies tradition and socio-technical studies has revealed how experiments spread by supplying both tangible evidence of impacts and outcomes and experiential evidence through the demonstration of alternatives in real-life settings (Marres 2009). Experimenting in cities promises scientifically rigorous knowledge that both reflects and is shaped by the context of lived experience and which as a result can be applied more quickly and successfully (Evans and Karvonen 2011). Various manifestations of the experimental mode of governance like living labs, maker spaces and hackathons hold the potential to reconnect the traditional political institutions of modernity – characterised by Beck as 'zombie' institutions that are dead but still alive (Boyne 2001) – with the experiences and needs of everyday urban life.

The promise of learning, and by extension innovation, lends experimentation considerable rhetorical power as a method through which to scale up from individual examples. As experimental activities reinterpret and reframe the trajectories of contemporary urban development, different frameworks are being developed to understand these processes. In their survey of Australian cities McGuirk *et al.* (2014) distinguish between institutional and practical experiments, with the former entailing experiments in arrangements within and between institutions to produce new ways of governing and the latter involving novel practical actions. Focusing on the transformative capacity of experiments, Smith and Raven (2012) distinguish between 'fit-and-conform' and 'stretch-and-transform' modes of change, which refer to experiments that take place within dominant institutional contexts versus those that transform their contexts. Practical experiments can prompt broader institutional change, and for many this represents the *sine qua non* of urban experimentation as a worthy pursuit. For others, urban experiments open up spaces for new kinds of governance and action in the city, giving centre stage to social interests that are downplayed under dominant governance arrangements but which over time may coalesce into coherent pathways to wider transformation (Bulkeley *et al.* 2015).

A key question emerging from the literature concerns the politics of experimentation, or more specifically who is allowed to take part at both the institutional and practical levels. For example, the smart city discourse as articulated in Europe and Asia has hitherto focused on trialling technological 'solutions' in real cities, privileging multinational corporations as urban actors (Vanolo 2014; Viitanen and Kingston 2014). In contrast, the Transition Towns movement positions local communities as the designers and instigators of urban experiments (Smith 2011; Seyfang and Haxeltine 2013, Aiken 2014, Feola and Nunes 2014). In reframing urban development experimentation shifts the balance of power between actors, empowering some while disempowering others, and privileging new forms of knowledge and evidence in the process (Karvonen *et al.* 2014). In some cases, the availability of information about the performance of experiments invites a data-driven approach to urban governance. In others, it militates a design-led approach to urban development, as cities and parts of them become positioned as urban living labs, serving as laboratories for radical change in which users are involved in the co-design of solutions to pressing urban problems (Nevens *et al.* 2013; Voytenko *et al.* 2015).

Much like the localism trap in development studies, which assumes that initiatives at the local level are somehow fairer, there is a tendency to assume that urban experimentation is an *a priori* beneficial endeavour. In their recent review, Luque-Ayala and Marvin (2015) highlight the need for smart urbanism to be more experimental, but Masdar City, perhaps the most high profile example of a smart eco-city, positions itself explicitly as an experimental city, albeit a highly techno-centric and corporatist one (Shelton *et al.* 2015). Experiments, understandings of experiments, and the attendant future visions they entail, are not inherently positive but carry politics just like any other development strategy.

Closely related and an increasingly important dimension of urban experimentation concerns how success is defined and measured. The quote at the start of this chapter is from Charles Kettering (1876–1958), an American inventor who led the research division of General Motors for 27 years. He patented the electric starter amongst numerous electric and lighting systems for cars, paving the way for the huge success of General Motors in establishing the automobile as a primary focus of urban planning in the second half of the twentieth century. The unintended consequences of this success – pollution, congestion, poor health and the destruction of communities – are both well-known and a central focus of cities that are striving to realise more sustainable urban futures. Strictly speaking, this example concerns innovation rather than experimentation, but the point holds in relation to how the evaluation of experiments can vary significantly depending on what outcomes are seen to constitute success. The definitions of success that inform evaluation often reflect the political goals and approaches of the specific actors involved, including researchers themselves (Voytenko *et al.* 2015). For example, transition scholars tend to highlight success when experimentation produces more environmentally friendly development pathways, while urban scholars highlight success when experiments are more socially and democratically robust. Caution is required here. Many different definitions of experimentation are at play, and the way in which experiments are designed, mobilised and evaluated differs hugely.

Debates concerning the politics of urban experimentation lead inexorably to the question of how an experiment or set of experiments drives wider transformation. This is a key topic of concern motivating this volume and many of the theoretical frameworks presented are essentially attempts to conceptualise this process. In relation to the low carbon agenda, numerous successful experiments have been established in cities over the last 20 years, leading funding bodies, policy makers, charities, companies and communities to a shared contemporary focus on how to translate discrete experiments into broader change. Part of the allure of experimentation is based on the assumption that it is possible to scale up from an individual project to the city through a process of trialling, learning and rolling out (Brown and Vergragt 2008; Evans 2011) but the complexity of achieving broader change is often hidden behind a lexicon of verbs such as upscaling, replicating, transforming, seeding, rolling out, and breaking through. These words imply quite different understandings of how change unfolds over space and time, blackboxing the social and political agency through which it takes place (Pesch 2015). While a revolution is enacted by revolutionaries, experiments, transformations and transitions have no obvious corresponding terms, despite the fact that they imply a power dynamic whereby certain (more powerful) groups are experimenting on other (less powerful) groups with the purpose of transforming or transitioning them.

Focusing on experiments directs attention to the specific social and material contexts in which urban change is embedded and through which it literally 'takes place'. Understanding experiments as sites through which 'particular urban infrastructure regimes … are configured and challenged' (Bulkeley *et al.* 2014: 1477) resonates with current understandings that emphasise the relational and provisional aspects through which the city is comprised (Graham and McFarlane 2015). Urban Political

Ecology with its emphasis on flows of power and materials, socio-technical studies with its emphasis on the coevolution of technology and society, and critical infrastructure studies focus on the ways in which urban institutions, techniques and artefacts are 'established, maintained and challenged' (Monstadt 2009: 14). The process of urban experimentation unfolds over space and time through reworking the relationships between social and material networks in the context of existing economic, social and political trajectories.

The body of work that has emerged around the idea of transitions presents an increasingly influential way to think about this relationship. Since its inception in the early 1990s, a bourgeoning literature on socio-technical sustainability transitions has drawn attention to how experimental approaches to innovation drive socio-technical change in the context of prevailing institutional, material and social structures (Markard *et al.* 2012). Drawing on evolutionary theories and socio-technical studies in particular, this literature is primarily concerned with understanding the emergence of new socio-technical systems and their interplay with system-level dynamics of incumbent ones. Incumbent socio-technical systems (or 'socio-technical regimes') are stabilised through the rigorous alignment of routines, institutions, infrastructures and networks that constitute the provision of societal needs such as energy, mobility and food. These regimes developed out of earlier responses to societal challenges and economic opportunities in the twentieth century, but now pose structural challenges to innovation responses to contemporary challenges such as climate change, global resource depletion, rapid urbanisation, and global economic restructuring. In evolutionary terms, regimes form disadvantageous selection environments for path-breaking innovations seeking to transform the very core of how we produce and consume goods and services. Successful experimentation as such entails not just short-term successes in experimental projects but more so their critical influence in transforming wider regime structures (Smith and Raven 2012).

Real-life experimentation with path-breaking innovations outside of laboratory spaces is a way to transcend this structural impasse (Kemp *et al.* 1998). It ideally enables reflexive and multi-dimensional learning processes across a range of issues such as infrastructures, policy paradigms, cultural norms, ways of organising markets, and consumer behaviour. As such, experimentation entails the negotiation of multiple expectations of possible and desirable futures (Berkhout 2006), and the re-making of social relations across chains of actors involved in or influenced by transitions in socio-technical regimes. The urban arena forms a relatively new context for socio-technical transitions research (Hodson and Marvin, 2010; Frantzeskaki *et al.* 2016), which historically drew national and sectoral boundaries around transitions in socio-technical systems (Raven *et al.* 2012). A key question is how urban and sectoral structures are entangled and form multi-scalar contexts for urban experimentation, including the role of power relations enacted through them (Murphy 2015; Truffer *et al.* 2015).

To summarise, urban experiments provide a unique point of intersection between practice and theory. They increasingly shape the activities of cities trying to transform themselves, as well as occupying the efforts of scholars from across a range of traditions to understand this process. By focusing on urban experimentation as a unit of study, this volume advances conceptual and practical understandings of the topic to identify common concerns, challenges and questions.

Contributions in this volume

The following chapters are contributions from leading scholars across the key disciplines currently at the cutting edge of the research field. They have a shared interest in understanding how experimentation is being conceived, implemented and assessed in a wide range of settings. The three questions of how urban experiments happen, who is involved, and what their wider impacts are provide recurring themes throughout their chapters. The book is organised into three parts: logics of urban experimentation, experimenting in cities, and experimental cities. The insights and examples that populate these three parts provide a framework to examine the phenomenon of urban experimentation in conceptual and empirical detail.

Part I, logics of urban experimentation, explores different theoretical and conceptual approaches to urban experimentation. It investigates the origins and undercurrents of urban thinking that resonate with and add to current debates about urban experimentation. Part II, experimenting in cities, presents a set of chapters that analyse how experiments are being staged within cities, focusing on who does the experimenting, on who or what, how and with what effects. This part reflects the wide variety of urban experiments that are currently in play, varying from buildings to neighbourhoods, and from highly formalised state-led experiments to informal attempts to remake the urban fabric. Part III, experimental cities, asks how entire cities or groups of cities are constructed and conceived as experiments and what can be learnt from understanding experimentation at the city level. This final part focuses on more iconoclastic examples of ambitious attempts to prompt rapid urban innovation through extreme forms of experimentation. As a note of clarification, the distinction between the terms urban experimentation and experimental cities is one of scale rather than process, reflecting a desire to explore both the content and context of urban experimentation. Parroting the distinction between urbanisation and the city offered by urban studies, urban experimentation can be viewed as a process that generates the experimental city as its outcome.

Part I: Logics of experimentation

This first part includes six chapters that examine different logics and theorisations of experimentation. The first chapter by Frans Sengers and colleagues reviews the notion of experimentation as discussed in the sustainability transitions literature. The chapter develops a typology of experimentation distinguishing between niche experiments, social experiments, transition experiments, grassroots experiments and sustainability experiments, and proposes an encompassing definition of experimentation. Tim May and Beth Perry critically examine how global dynamics, in particular those related to neoliberal capitalism and the knowledge economy, are framing the conditions within which experimentation takes place. They discuss the possible dangers of such a framing, in which neoliberal practices do not positively benefit city populations but rather risk accelerating the inequalities produced by business-as-usual approaches. They propose an alternative framing around the notion of grassroots experimentalism. Simon Marvin and Jonathan Silver draw on an extensive survey of over 70 urban laboratories to develop a typology of multiple styles of experimentation. The typology captures the huge variety in terms of the focus, setting, logics, activities, and temporal orientation of experimentation in urban laboratories. Reflecting on the results, they call for more comparative research to better capture the political roles and social organisation of urban laboratories.

In their chapter, Chris Ryan and colleagues report on recent attempts to engage in experimentation from a visual design perspective. They build upon collective envisioning and participatory methods promoted in Transition Management approaches and combine these with recent design techniques. The resulting process is organised around debating 'glimpses of possible futures' in interactive workshop settings. Although it is too early to make any definitive conclusions about how this process contributes to actual transformations on the ground, the preliminary results illustrate that visual design experimentation enables a necessary cognitive break from the status quo. Elizabeth Rapoport looks at the recent history of 'sustainable urban projects' – large-scale efforts to create a new district or area of a city with explicit attempts to integrate sustainable design criteria. She highlights how branding and marketing, and in particular voluntary rating and certification schemes, are the most common reason to incorporate sustainability in these projects. She also argues that radical innovation is unlikely in a constrained space of urban development due to risk aversion amongst the rather narrow group of sustainable property developers involved. Simon Marvin and Mike Hodson end this section with a chapter on 'cabin ecologies' – enclosed life support systems. They trace the intriguing history of the development and deployment of integrated urban infrastructures from its origins in space and military programmes, and reflect on the path dependencies and implications of these developments for more recent attempts such as ecological urbanism and urban control rooms.

Part II: Experimenting in cities

Part II provides different interpretations of how experiments are situated in particular material and institutional contexts. Martin Sanzana Calvet and Vanesa Castán Broto explore the emergence of green enclaves in peri-urban areas of Santiago, Chile through state and market-led experiments. Applying ideas from Urban Political Ecology, they argue that green enclaves reveal the contradictory character of experiments as simultaneously emancipatory while reproducing and facilitating the existing neoliberal political agenda. Their chapter highlights the connections between greening and social power that have significant political implications for urban experiments. Duke Ghosh and colleagues focus on mobility experiments in India and Thailand using ideas from socio-technical transitions theory. They follow the actors in four case studies to reveal the various navigational strategies employed to realise systemic change in incumbent and obdurate transportation networks. Their analysis provides a comparative framework for understanding the interplay between context and strategies of experimentation.

Gareth Powells and Lynsay Blake explore experimentation in a very different way in Newcastle upon Tyne in northeast England. They focus on the dynamics of an urban science network comprised of actors from the private, public and third sectors, and highlight the growing role of universities in local partnerships in generating new knowledge about sustainable urban development. The particular forms of experimentation that emerge are negotiated and shared amongst the various partners in an attempt to align their agendas. Jana Wendler argues that experiments are not always intentionally designed but can emerge organically over time. Her study of a community garden in Berlin demonstrates how everyday activities can produce an arena of learning and improvisation that is distinct from more formalised experimental activities. These grassroots modes of innovation provide opportunities for more creative and open-ended forms of experimentation and suggest a latent potential in social movements to influence urban trajectories. Finally, Gabriele Schliwa and Kes McCormick focus on the increasing use of living laboratories in European cities to address sustainability challenges, providing a comprehensive review of the literature in this field. Drawing on case studies they identify a clear distinction between living labs that are user-centric and those that are citizen-centric, associating the latter with a focus on social learning and a greater capacity to prompt lasting transformation.

Part III: Experimental cities

Part III comprises five chapters that explore how entire cities or groups of cities are constructed and conceived as experiments. Sofia Shwayri contrasts two case studies that reveal the diversity of experimentation characterising current urban planning in Seoul, as well as how the city has learnt from other experiments. The chapter situates both accounts within the wider trajectory of Korean development and hints at a specifically Korean approach to urban experimentation based on the 'bali bali' philosophy. Federico Cugurullo uses Masdar City, a well-known experimental eco-city project under development in the United Arab Emirates, to demonstrate the fragmentary character of urban experiments. Masdar City is understood as an urban experiment composed of multiple sub-experiments such as smart grid installations, large-scale pedestrian spaces, fusions of vernacular and modern architectural styles, and automated transport systems. The number of actors involved, divergent understandings of sustainability and fragmented projects ultimately prevents Masdar City from achieving its vision of the sustainable city, revealing fundamental tensions between urban experimentation and sustainability.

Thomas Yarrow offers a compelling post-mortem of a past urban experiment, focusing on a mid-century failed experiment in urban living in Ghana. Informed by the tenets of modernist planning, the Volta Resettlement Project offers insights into the challenges of experimenting on the city scale. The chapter explores the relationships between social experiments, forms of expertise, and the emergence of a specifically urban mode of development. Yarrow also discusses how the alternative futures promised by experiments are reconceived after their failure, arguing that far from revelling in its subversion the resultant urban ruins prompt fidelity to the original vision. In the penultimate chapter, James Evans and colleagues explore two cases that have explicitly positioned themselves as experimental cities: Arcosanti, founded in the 1960s in Arizona, and Masdar City, a current-day attempt to discover the blueprint for sustainable urbanism in the Middle Eastern desert. The authors reflect on the relation between the design and reality of these experimental cities, showing how contrasting efforts have run up against similar obstacles in practice and arguing that urban experiments need to take place within a functioning political community if they are to achieve broader impacts. In the final chapter, Stephanie Pincetl considers what a post carbon city might look like at street level, discussing the broader legal and regulatory challenges that hinder the widespread transformation of urban areas. This forthright think-piece poses some of the bigger, 'harder to answer' types of questions that lurk behind contemporary discussions of urban experiments and sustainability. The chapter broadens the purview of the book, setting the experimental endeavour within the context of deeper urban transformation and what might have to happen to enable fundamental change.

Towards experimental urbanism?

Urban experiments are intriguing because they constitute explicit attempts to stage and learn about different possible futures in the real world; their actuality matters in producing a different kind of city. They offer novel modes of engagement, governance and politics that both challenge and complement conventional strategies in important ways. Taken as a whole, the chapters demonstrate the broad set of impacts that experimentation is having on cities and processes of urban development. They show how experimentation is increasingly informing urban practices, reshaping understandings of the city and the related practices of knowledge production that sustain it across a wide variety of sectors, including carbon governance, energy services, technological innovation, transportation networks, economic growth, and social organisation.

The incorporation of urban experiments into the mainstream of urban planning and thought is producing a nascent, stabilised set of precepts that could be termed 'experimental urbanisation'. This entails a dynamic, yet highly political, mode of governance that is continually experiencing change as the result of targeted interventions. The contributions to this volume indicate a growing effort to situate urban experimentation as a mode of governance within broader understandings of the material and political production and reproduction of cities and parts of cities (Merrifield 2014; Brenner and Schmid 2015; Walker 2015). The maturation of this field of study is characterised by a deeper engagement with and dialogue between the theoretical canons of urban studies and socio-technical transition studies to flesh out the dimensions of a specifically experimental mode of urbanisation.

The contributions in this volume reveal how the current round of urban experimentation differs from previous incarnations, representing a specific kind of governance 'fix' for a broadly neoliberal system that is struggling to move towards more sustainable forms of urban development. As a distinct mode of urban governance, the challenge of experimentation is to bring people and infrastructures together in ways that are capable of realising significantly different and more sustainable urban futures in socially just and democratic ways (Evans and Karvonen 2014). A key question raised by the research represented in this volume is whether the experimental mode constitutes a subversive or reinforcing element of contemporary urbanisation. Many of the contributors highlight the implicit assumptions that experimental urbanism carries about what a city is and who should be involved in creating and managing it. The tendency for experimental approaches to be captured by dominant interests and produce socially and politically fragmented cities is clear, and provides an important corrective to the sometimes overly exuberant and uncritical celebration of urban experimentation among academics and policy makers. Clearly, urban experimentation is neither entirely subversive nor entirely reinforcing of the status quo, but includes activities that are one or the other and sometimes both. There is a clear need here to understand how different modes of experimentation coexist and interrelate at larger urban and regional scales.

Related to this, the contributions indicate that the current research agenda in this field is increasingly focusing on how to embed experimentation into cities in long-term and more meaningful ways, paying attention both to the micro-scale social and political practices, impacts, and implications of experimentation, as well as to the larger-scale networks and policies that sustain them. These concerns resonate with those of practitioners and policy makers, who are increasingly focused on moving past isolated experiments to consider how more long-term and varied modes of experimentation can stimulate broader urban transformation. Understanding the durability and multiplicity of experiments within their broader urban context is a necessary first step towards recuperating experimental urbanism as a progressive driver of change. Given the current explosion of urban experimentation around the world, it is hard to imagine a more important and timely research agenda.

References

Aiken, G. (2014). (Local-) community for global challenges: carbon conversations, Transition Towns and governmental elisions. *Local Environment*, 20: 764–781.

Beck, U. (1995). *Ecological Enlightenment: Essays on the Politics of the Risk Society*. Atlantic Highlands, NJ: Humanities Press.

Berkhout, F. (2006). Normative expectations in systems innovation. *Technology Analysis and Strategic Management*, 18: 299–311.

Boyne, R. (2001). Cosmopolis and risk: a conversation with Ulrich Beck. *Theory, Culture and Society*, 18(4): 47–63.

Brenner, N. and Schmid, C. (2015). Towards a new epistemology of the urban? *City*, 19: 151–182.

Brown, H. and Vergragt, P. (2008). Bounded socio-technical experiments as agents of systemic change: the case of zero-energy residential building. *Technological Forecasting and Social Change*, 75: 107–130.

Bulkeley, H. and Castán Broto, V. (2012). Government by experiment? Global cities and the governing of climate change. *Transactions of the Institute of British Geographers*, 37: 1–15.

Bulkeley, H., Castán Broto, V. and Maassen, A. (2014). Low-carbon transitions and the reconfiguration of urban infrastructure. *Urban Studies*, 51: 1471–1486.

Bulkeley, H., Castán Broto, V. and Edwards, G. (2015). *An Urban Politics of Climate Change: Experimentation and the Governing of Socio-Technical Transitions*. London: Routledge.

Castán Broto, V. and Bulkeley, H. (2013). A survey of urban climate change experiments in 100 cities. *Global Environmental Change*, 23: 92–102.

de Jong, M., Joss, S., Schraven, D., Zhan, C. and Weijnen, M. (2015). Sustainable-smart-resilient-low carbon-eco-knowledge cities; making sense of a multitude of concepts promoting sustainable urbanization. *Journal of Cleaner Production*, 16: 25–38.

Evans, J. (2011). Resilience, ecology and adaptation in the experimental city. *Transactions of the Institute of British Geographers*, 36: 223–237.

Evans, J. and Karvonen, A. (2011). Living laboratories for sustainability: exploring the politics and epistemology of urban adaptation. In Bulkeley, H., Castán Broto, V., Hodson, M. and Marvin, S. (eds), *Cities and Low Carbon Transitions*. London: Routledge, 126–141.

Evans, J. and Karvonen, A. (2014). Give me a laboratory and I will lower your carbon footprint!' – Urban laboratories and the pursuit of low carbon futures. *International Journal of Urban and Regional Research*, 38: 413–430.

Feola, G. and Nunes, R. (2014). Success and failure of grassroots innovations for addressing climate change: the case of the Transition Movement. *Global Environmental Change*, 24: 232–250.

Frantzeskaki, N., Castán Broto, V., Coenen, L. and Loorbach, D. (eds) (2016). *Urban Sustainability Transitions*. London: Routledge.

Graham, S. and McFarlane, C. (eds) (2015). *Infrastructural Lives: Urban Infrastructure in Context*. London: Routledge.

Gross, M. (2010). *Ignorance and Surprise: Science, Society and Ecological Design*. Cambridge, MA: MIT Press.

Hodson, M. and Marvin, S. (2010). Can cities shape socio-technical transitions and how would we know if they were? *Research Policy*, 39: 477–485.

Karvonen, A. and van Heur, B. (2014). Urban laboratories: experiments in reworking cities. *International Journal of Urban and Regional Research*, 38: 379–392.

Karvonen, A., Evans, J. and van Heur, B. (2014). The politics of urban experiments: radical change or business as usual? In Marvin, S. and Hodson, M. (eds), *After Sustainable Cities*. London: Routledge, 105–114.

Kemp, R., Schot, J.W. and Hoogma, R. (1998). Regime shifts to sustainability through processes of niche formation: the approach of strategic niche management. *Technology Analysis and Strategic Management*, 10: 175–198.

Luque-Ayala, A. and Marvin, S. (2015). Developing a critical understanding of smart urbanism? *Urban Studies*, 2: 2105–2116.

McFarlane, C. (2011a). *Learning the City: Knowledge and Translocal Assemblage*. London: John Wiley & Sons.

McFarlane, C. (2011b). The city as a machine for learning. *Transactions of the Institute of British Geographers*, 36: 360–376.

McGuirk, P., Dowling, R. and Bulkeley, H. (2014). Repositioning urban governments? Energy efficiency and Australia's changing climate and energy governance regimes. *Urban Studies*, 51: 2717–2734.

Markard, J., Raven, R.P.J.M. and Truffer, B. (2012). Sustainability transitions: an emerging field of research and its prospects. *Research Policy*, 41: 955–967.

Marres, N. (2009). Testing powers of engagement: green living experiments, the ontological turn and the undoability of involvement. *European Journal of Social Theory*, 12: 117–133.

Merrifield, A. (2014). *The New Urban Question*. London: Pluto.

Monstadt, J. (2009). Conceptualizing the political ecology of urban infrastructures: insights from technology and urban studies. *Environment and Planning A*, 41: 1924–1942.

Montgomery, C. (2013). *Happy City: Transforming Our Lives Through Urban Design*. London: Penguin.

Murphy, J. (2015). Human geography and socio-technical transition studies: promising intersections. *Environmental Innovation and Societal Transitions*, 17: 73–91.

Nevens, F., Frantzeskaki, N., Gorissen, L. and Loorbach, D. (2013). Urban transition labs: co-creating transformative action. *Journal of Cleaner Production*, 50: 111–122.

Pesch, U. (2015). Tracing discursive space: agency and change in sustainability transitions. *Technological Forecasting and Social Change*, 90: 379–388.

Raven, R.P.J.M., Schot, J.W. and Berkhout, F. (2012). Space and scale in socio-technical transitions. *Environmental Innovation and Societal Transitions*, 4: 63–78.

Seyfang, G. and Haxeltine, A. (2013). Growing grassroots innovations: exploring the role of community-based initiatives in governing energy transitions. *Environment and Planning C: Government and Policy*, 30: 381–400.

Shelton, T., Zook, M. and Wiig, A. (2015). The 'actually existing smart city'. *Cambridge Journal of Regions, Economy and Society*, 8: 13–25.

Smith, A. (2011). Community-led urban transitions and resilience: performing Transition Towns in a city. In Bulkeley, H., Castán Broto, V., Hodson, M. and Marvin, S. (eds), *Cities and Low Carbon Transitions*. London: Routledge, 159–177.

Smith, A. and Raven, R.P.J.M. (2012). What is protective space? Reconsidering niches in transitions to sustainability. *Research Policy* 41: 1025–1036.

Truffer, B., Murphy, J. and Raven, R.P.J.M. (2015). The geography of sustainability transitions. Contours of a quickly emerging research field. *Environmental Innovation and Societal Transitions*, earlyview online.

Vanolo, A. (2014). Smartmentality: the smart city as disciplinary strategy. *Urban Studies*, 51: 883–898.

Viitanen, J. and Kingston, R. (2014). Smart cities and green growth: outsourcing democratic and environmental resilience to the global technology sector. *Environment and Planning A*, 46: 803–819.

Voytenko, Y., McCormick, K., Evans, J. and Schliwa, G. (2015). Urban living labs for sustainability and low carbon cities in Europe: towards a research agenda. *Journal of Cleaner Production*, earlyview online.

Walker, R. (2015). Building a better theory of the urban: a response to 'Towards a new epistemology of the urban?' *City*, 19: 183–191.

Wieczorek, A., Raven, R.P.J.M. and Berkhout, F. (2015). Transnational linkages in sustainability experiments: a typology and the case of solar photovoltaic energy in India. *Environmental Innovation and Societal Transitions*, 17: 149–165.

Part I

Logics of experimentation

2 Experimenting in the city

Unpacking notions of experimentation for sustainability

Frans Sengers, Frans Berkhout, Anna J. Wieczorek and Rob Raven

Introduction

This chapter articulates a comprehensive definition of the notion of 'experimentation' in the context of urban sustainability transitions. Highlighting a number of key dimensions, we argue that experiments can be analysed according to the degree to which they are (1) inclusive, (2) systemic, (3) practice-based, (4) challenge-led, (5) sites of social learning and (6) adaptive in the face of uncertainty and ambiguity in fostering socio-technical change. We point out a number of urban governance challenges related to coordination, commitment and negotiation, and we highlight a number of promising avenues for future research on experimenting in the city.

The notion of experimentation occupies a central position within the field of sustainability transitions.[1] It is this focus that sets the sustainability transitions field apart from the wider literature on social change and policy theory (Meadowcroft 2011; van den Bergh 2012). Experiments represent important seeds of change that may eventually lead to profound shifts in the way societal functions such as the provision of energy or mobility are met. As precious yet-to-germinate microcosms of sustainable systems and practices, the alternative socio-technical configurations embodied in experiments are applied and tested in real-life contexts with the aim of technological, social and institutional learning. The promise is that learning and demonstration effects of experiments add to the momentum of emerging sustainable configurations which are geared to transform unsustainable socio-technical systems.

However, within sustainability transitions research, there are several different and productive ways to frame experimentation. For instance, the approach of Strategic Niche Management (Kemp *et al.* 1998) stresses that practical, hands-on experiments are the starting point of a transition pathway, while the Transition Management approach (Rotmans *et al.* 2001) proposes that actors should engage in efforts of collective envisioning before agreeing on the set-up of an experimental project. Other work on experimentation highlights the role of civic engagement in local communities (Seyfang and Haxeltine 2012) or frames experiments as a basis for innovation for lower-income economies to embark on green growth trajectories (Berkhout *et al.* 2010). Experimentation has also been seen as a response to stagnation in conventional policy approaches and as part of the broader trend of the fragmentation of vested authority that creates spaces for new sources of authority, legitimacy and action by new social actors (Hoffman 2011). In short, in the study of experimentation there is a wide variety of goals and values, theoretical underpinnings and discursive emphasis, actors and places.

Interest in the role of cities as arenas for experiments for system innovations linked to sustainability transitions stands in a long tradition of scholarly concern with the economics of agglomeration. Classically, the benefits of agglomeration of business and innovative activity in cities has been associated with physical accessibility, the concentration of physical capital and infrastructures, and the availability of professional labour (Marshall 1920). In contemporary economic geography, there is a widely held view that the local urban 'milieu', involving clusters of small firms in flexible alliances benefitting from mutual information exchanges and associated information spillovers, provides the basis for successful innovation. Gordon and McCann (2005: 528) argue that the more intense creativity and entrepreneurship seen in cities is due to:

- i a rich 'soup' of skills, ideas, technologies, and cultures within which new compounds and forms of life can emerge;
- ii a permissive environment enabling unconventional initiatives to be brought to the marketplace; and
- iii vigorously competitive and critical arenas operating selection criteria which anticipate (and/or shape) those of wider future markets.

More recently, geographers have brought to the innovation studies-inspired field of sustainability transitions a range of new terms and ways of thinking about urban experiments conducted in contemporary cities (Evans and Karvonen 2011; Bulkeley *et al.* 2015).

In an earlier workshop contribution, we presented a review of how transition scholars have made sense of the many urban sustainability initiatives and projects being conducted today (Sengers *et al.* 2014). In a systematic literature review, we distinguished different types of experiments and some trends that emerge from this research.[2] In this chapter, we build on these efforts by deriving a comprehensive definition of experimentation. We mobilize our definition to point out a number of governance challenges for experimenting in the city and, finally, we highlight a number of promising avenues for future research.

Experimenting in the transitions literature

Transition scholars have proposed many terms for describing experimental projects. Taken together, these definitions make a distinction between the initiatives in the context of transitions and the idea of an experiment in scientific research. Experiments in research, whether in natural or social science, are procedures designed to establish the relationship between phenomena by controlling their interaction. The researcher designs the set-up of the experiment and aims to control all relevant aspects of the process as far as possible separated from the complexity of real-world conditions. In the experiments that transition scholars are interested in, there is still an interest in designed procedures, but the degree of separation from the world and the degree of control over relevant conditions will generally be much lower than in a scientific experiment. Contrary to scientific experiments, the experiments conducted in the context of transitions are not designed to establish facts about a single causal relationship but aim to simulate a complex process of social and technological co-evolution with emergent properties

Experiments in the context of transitions are also distinct from the idea of 'development' as captured in the term research-and-development (R&D), although they share the emphasis on learning-by-doing. Development in the classical sense is usually related to the activities of industrial laboratories (complementing scientific experiments) and relates to the testing and refining of new technologies and practices under controlled conditions. In development, the conditions of working gradually come to resemble the real-world conditions under which the technology will come to operate. Development covers a broad set of activities (Cohen 2010), often including the collaboration amongst a technology-holding firm, customer firms and end users. The primary aim is to develop proprietary, appropriable knowledge and know-how for commercial advantage for the parties involved. The distinction with transition-oriented experiments is primarily related to the greater diversity of social actors involved and the broader ambition in learning from the experiment. Typically, initiatives described as experiments in the context of transitions will involve a broad coalition of parties (firms, end users, civil society organizations, research institutes, regulators, government agencies, and so on) with the objective of connected social learning in relation to a new socio-technical configuration, and with the aim of generating both proprietary and shared, public knowledge.

Much of the work in the conceptual framing of experiments in the context of transitions has gone into clarifying who participates, what is learned, and who appropriates what has been learned. These are important questions because they point to the incentives and risks that social actors face when participating in experiments, to the likely sources of resources that will support these initiatives, and to the distribution of benefits and costs that will be generated. Below is an overview of how experiments feature in the transitions literature.

The main type of experiment described in the transitions literature – and a root of much of the current research in the field of sustainability transitions – is the 'niche experiment' (Kemp *et al.* 1998). Niche experimentation is a central idea in the Strategic Niche Management (SNM) literature, which developed from the observation that there is an abundance of promising new technologies being developed by firms and in private–public arenas, many of which fail to develop into widely adopted innovations. The diagnosis of this pattern of innovation was that mature socio-technical regimes – stable, well-ordered, path-dependent socio-technical configurations – are geared toward incremental innovation and select against radical novelties (Kemp *et al.* 1998; Geels 2002). SNM assumes that the creation of new niches through protective policy measures and strategic development of existing market niches is needed to shift from a dominant, but unsustainable, socio-technical regime to alternative, more sustainable regimes and pathways (Schot and Geels 2008). Niches provide protective loci for the development of path-breaking innovations and empower experimental projects (Smith and Raven 2012) so that they can be 'up-scaled' (Jolly *et al.* 2012), disrupting and transforming dominant regimes.

Niches are protected spaces in which experiments can take place away from prevailing selection pressures (whether in a market or not) that would otherwise be damaging to them. 'Outsiders' to the incumbent regime are, next to users of innovations, considered critically important actors (van de Poel 2000). Niche experiments offer protection through three processes: 'shielding' to hold off the regime-defined selection pressures; 'nurturing' to support technological learning, business development and the emergence of an institutional framework supporting innovation; and 'empowering' to make niche innovations competitive vis-à-vis regimes (Smith and Raven 2012).

Research on 'bounded socio-technical experiments' followed from earlier work on 'social experiments' (Verheul and Vergragt 1995). This concept surfaced as a critique of SNM's focus on creating a market for new technology. The social experiment emphasizes the importance of social innovation and the role of civil society, citizens and consumers. It highlights the process of building a social network of diverse actors promoting practices, institutions and norms that are at the basis of novel socio-technical regimes. This core idea was further developed into the 'bounded socio-technical experiment' (BSTE). The BSTE attempts to introduce a new product or service on a scale bounded in place and time within a specific community. In contrast to the niche experiment, which is rooted in evolutionary thinking, the BSTE draws on social learning literature (Bandura 1977; Argyris and Schön 1978) and the process of learning needs to be organized and monitored as a part of action research (Brown and Vergragt 2008).

Contrary to previous work on niche experiments and BSTEs, 'transition experiments' do not necessarily feature technological change or environmental sustainability. They are innovation projects that explore radically new ways of meeting societal needs and solving persistent societal problems (van den Bosch and Rotmans 2008). Transition experiments are framed in the context of complex systems and evolutionary theory-based Transition Management (Rotmans *et al.* 2001; Loorbach 2007). Transition Management aims to influence and modulate (sustainability) transitions. Although it sets out to motivate structural change, the evolving goals of a transition is the outcome of an inclusive and participatory process between diverse participants in a transition experiment. The aim is to build on the existing social and economic changes and interactions to re-orient processes of structural change to goals actively chosen by a wide range of societal actors. The analytic emphasis in Transition Management is on three processes: 'deepening' to learn within the experiment; 'broadening' to learning from related experiments; and 'scaling-up' to learn about regime change and broader developments (van den Bosch 2010). 'Frontrunners' are viewed as critical agents for conducting transition experiments. They gather in a transition arena, the outcome of which includes a portfolio of transition experiments that codify and elaborate the developed societal vision.

Motivated by social activism, 'grassroots experiments' refer to networks of activists and organizations generating novel bottom-up solutions that respond to local social and cultural contexts, and the interests and values of the communities involved. Grassroots initiatives operate in civil society arenas (e.g. cooperatives, voluntary associations or social enterprises) that experiment with social innovations and greener technologies (Seyfang and Smith 2007). Their innovative potential can be evaluated in relation to their intrinsic benefits (when an experiment is valued for its own sake and not geared toward regime change), their diffusion benefits (when the aim is to contribute to wider transformation), and the extent to which they promote measurable changes in the sustainability of consumption (Seyfang 2010).

'Sustainability experiments' have been defined as 'planned initiatives that embody a highly novel socio-technical configuration likely to lead to substantial (environmental) sustainability gains' (Berkhout *et al.* 2010: 262). The concept emerged in the context of debate about alternative development pathways in developing Asia and sought to draw together insights from growth theory with insights from research about the resource and environmental intensity of economies (Berkhout *et al.* 2009, 2010, 2011). Conventional economic growth research argues that business firms in latecomer countries need to develop technological capabilities at the global technological frontier, usually by becoming competitive through trade with more demanding markets in advanced economies, before they are able to move from imitation to innovation (Kim 1997). The capacity to innovate is difficult for firms to acquire and will first appear at later stages of economic development, and initially in advanced sectors of the economy. This suggests that innovations underpinning more resource-efficient development pathways are also unlikely to emerge in developing economies but will continue to be transferred from firms in advanced economies.

Recent research in developing Asia has identified a great number of small-scale innovative projects for sustainability – from solar home systems (Wieczorek *et al.* 2015) to new forms of mass transit (Sengers and Raven 2015) – which appear to confound this picture. One explanation is that they bring together capabilities not just of business firms, but also of civil society, and that they draw heavily on transnational flows of expertise, technology, capital and institutions to innovate socio-technical configurations that are better aligned with the market and cultural contexts in lower-income countries (Sengers and Raven 2015; Wieczorek *et al.* 2015). These 'transnational linkages' articulate, complement and motivate local capability formation and may contribute to alternative development pathways that defy traditional growth theories, offering 'green growth' pathways to development.

Although this typology of experiments describes an evolution of scholarly thought about experimentation in the transitions literature, it is not an exhaustive list. Other terms include 'governance experiments' (Bos and Brown 2012), highlighting the configuration of decision-making that emphasizes collaborative planning, participation and social learning as a way to look beyond the disproportionate focus on technical experimentation in SNM, and 'real-world experiments' (Schneidewind 2012), which highlight urban projects that involve the public in ecological restoration. Another important new term is the 'urban climate change experiment' (Castán Broto and Bulkeley 2013; Bulkeley *et al.* 2015) which draws on Foucault's work on 'governmentality', combined with ideas from transition studies, STS (Science and Technology Studies), policy studies and urban political ecology. Bulkeley and colleagues (2015) provide a large sample of 'interventions' where climate change is 'put to work' locally in a variety of urban settings. These become climate change experiments when they are (1) purposive and strategic, (2) geared toward the mitigation of or adaptation to climate variability and change, and (3) delivered in the name of an urban community.[3]

Bringing together the strands: toward a definition of experimentation in sustainability transitions

Tracing the different conceptualizations of experimentation through time, we contend that two main lines of thought have emerged. The first line follows the initial work on niche experimentation, geared toward the creation of markets in which new green technologies are widely adopted and come to reshape socio-technical regimes. Perhaps because Strategic Niche Management initially drew on Constructive Technology Assessment as a way to analyse critically and support the guided experimental introduction of new technologies in society, it is today often perceived as putting too much emphasis on technological innovation. A second line, developed in response, has emphasized social innovation by networks of social actors from business, civil society and government. While the first line underpins the notions of niche experiment and sustainability experiment, the second line is apparent in the emergence of terms such as bounded socio-technical experiment and grassroots experiment.[4]

While we have presented a stylized typology, the distinctions between types of experiments are actually much more subtle and fluid. Indeed, many real-life transition projects described in the literature as one type of experiment could also be described in the terms of other experiment types.[5] We have chosen to highlight the differences to reveal the diversity and scope of experimental transition projects undertaken as well as the multiplicity of ways of making sense of these interventions. Following the evolution of research carried out under the banner of the experiment, we see an expansion in both the scale and scope of the projects studied. While earlier work emphasized technological innovation and state-or-firm-driven experimentation in OECD (Organisation for Economic Co-operation and Development) countries, more recent work has included a larger variety of projects, sometimes in the Global South and often foregrounding social innovation by engaged citizens.

It is also clear that more recent work in the field of sustainability transitions has tended to focus on urban experiments, which are then analysed at the scale of the city or city-neighbourhood. Many recent examples attest to this: green urban transport innovations in Göteborg, Hamburg and Curitiba (Carvalho *et al.* 2012); sustainable housing in Bangalore, Sao Paulo and Philadelphia (Bulkeley *et al.* 2015); and new types of bus-based mass transit in Bogotá, Bangkok and beyond (Sengers and Raven 2015). As opposed to earlier work on transitions, which conceived of experiments and niches as national-level entities, these cases show that cities – often connected to other cities across the globe – are important in their own right as arenas of experimentation and as drivers of sustainable innovation (also see Bulkeley *et al.* 2011).

Despite these differences, oppositions and shifting trends, there is much that binds thinking about transition-oriented experimentation and a comprehensive definition can be formulated. Within scientific practice, the term 'experiment' is often narrowly interpreted as a planned investigation carried out to test a hypothesis by providing insight into causes and effects by testing outcomes when particular factors are manipulated. But the term also connotes learning through 'experience' by trialling something, '… a course of action tentatively adopted without being sure of the outcome' (Oxford Dictionaries 2015). As we have seen, in the literature the experiment has a rather specific set of meanings. But as we have also seen, there is plenty of conceptual diversity and flexibility. Taking into account differences as well as common ground, we define an experiment in the context of sustainability transitions as:

> An inclusive, practice-based and challenge-led initiative designed to promote system innovation through social learning under conditions of uncertainty and ambiguity.

The notion of *system innovation* in this definition refers to co-evolutionary processes of far-reaching changes in technological, institutional, infrastructural and social dimensions of a system (for instance, a city or an energy system). For example, a famous experiment with an innovative type of bus-based transport system in the Brazilian city of Curitiba was discursively, institutionally and infrastructurally geared to undermine the reproduction of the unsustainable automobility regime in favour of an alternative multi-modal transport system (Sengers and Raven 2015)

Given this long-term and largely normative context, the transitions literature generally argues that experiments need to be *inclusive* of a variety of engaged social actors. Experiments encompass diverse actors – firms, users, civil society organizations, research institutes, regulators, and government agencies. New forms of partnerships may be involved, with hybrid relationships – commercial, public–private and informal – emerging. Inclusion aims to stimulate learning about the range of impacts, benefits, costs and risks for different social actors generated through system innovation. By identifying cultural, ethical and distributional objections early, enabling democratic decision-making through participation, and making refinements in response, there is a greater chance that these innovations are 'socially-robust' and fair (Gibbons 1999). The distribution of expected value and risk amongst participating actors is of key concern here. In principle, a broader range of participants and a more open and collaborative process of learning will result in an innovation with better alignment of diverse interests and values and greater social distribution of benefits. For example, at a time when the plying of rickshaws through India's cities was being criticized for being a profession that is 'inhumane' for the operators, the idea of electrifying the cycle rickshaw emerged as an option that could be supported by a wide range of actors, who collectively facilitated an experiment with these new vehicles on the streets of New Delhi (see Chapter 9 in this volume).

Experiments are *practice-based* interventions because they occur outside of the conventional laboratory under real-life conditions. For example, after a new type of portable taxi-meter for motorcycle taxis was developed under controlled conditions, it was exposed to physical and social reality when it was trialled and tested for real-life use on the streets of Bangkok (Sengers and Raven 2014). Such interventions are not just concerned with technological and economic performance, but with broader social performance (however defined) of the chains of connected innovations, adjustments and realignments that unfold through the experiment. Under such real-life conditions, the ability of profit-seeking actors (businesses) to appropriate knowledge to achieve competitive advantage over rivals in the market will be at risk. Hence, the willingness of businesses to invest in innovation under such conditions can be expected to be low. Given the unconventional appropriability conditions that surround experiments, we can expect an expanded role for public support for longer periods than is usually assumed in conventional innovation theory, where public support is withdrawn as technological innovations come closer to commercialization and firms are able to secure 'innovation rents'. Calibrating public support under these conditions is likely to prove to be a major governance challenge.

Experiments are sites of *social learning* (Brown *et al.* 2003) that extend beyond narrow techno-economic assessments.[6] Technological learning in the conventional sense of refining technical performance and bringing down costs is one aim of experiments. But beyond this, an experiment also aims to stimulate practical, network and institutional learning, including the development of new business models, producer–user relationships, regulations, user and social practices and so on that may follow from and constitute system innovation. Facilitating this kind of 'systemic learning-by-doing' requires a broad-based and iterative approach to learning, with inclusive learning goals, and monitoring and evaluation of progress across a series of steps and phases. For example, while the aforementioned experiment with the motorcycle taxi-meter provided new information about technical real-life performance of the new device, it also provided the initiators with a rich learning experience in terms of dealing with other stakeholders and navigating the complex institutional environment of a semi-formal taxi industry (Sengers and Raven 2014).

Another distinctive characteristic of experiments is the co-design of the problem frame by the diverse network of social actors involved, summarized in the notion of a *societal challenge*. Conventionally, innovative activities within business firms are organized against the background of a commonly accepted techno-economic paradigm and with agreement about the specific problems to be addressed (Dosi 1982; Nelson and Winter 1982). With the problem frame established, the firm decides for itself how to deploy its assets and capabilities to address specific opportunities in the context of market competition and regulation. The firm will test innovations with potential users and customers but this will generally be under conditions that protect intellectual property and know-how, so enabling Schumpeterian innovation rents to be collected later. Purposive system innovation – and the experiments it builds from – typically starts with a societal challenge for which prevailing techno-economic paradigms and socio-technical regimes are widely believed not to provide satisfactory solutions. For example, the NGO (non-governmental organization) AuroRE set up experiments with solar home systems in India as a way to explicitly address multiple societal challenges such as deep reductions in CO_2 emissions and the need for affordable lighting systems for marginalised populations (Wieczorek *et al.* 2015).

The broader socio-technical problem frame is itself in question and its renegotiation is part of the work of developing new socio-technical configurations in experiments. The process of experimenting for sustainability transitions can be seen as a collective design process, involving a broad range of social actors in search of a new problem frame, performance metrics and a functional configuration of technologies and practices to generate desired goods and services. But this broadened social basis for innovative activities – beyond businesses, their supply chains and the customers, and concerned with reconfiguring a techno-economic problem frame and solving the complex problems that emerge from systems innovation – also implies new problems and costs. The inclusion of a greater variety of social actors and interests implies that there will be diverse – and often misaligned and conflicting – values and interests at play. The negotiation of these values through an experiment is an intrinsic part of the process of innovation and social learning, introducing *ambiguity* to goal-setting and problem-solving. This also introduces deep *uncertainty* into the pace, direction and substance of the problems, obstacles, solutions, adjustments and changes that emerge and are tackled through the experiment. We draw a distinction here between the depth and dimensionality of the uncertainties implied by socio-technical experiments and the degree and nature of uncertainty common to more conventional demonstration projects. The design of experiments therefore needs to be adaptive to ambiguity and uncertainty. For example, the innovative bus-based transport system in the city of Ahmedabad is hailed by some actors as an experiment that embodies socially inclusive mobility for all and as an emerging socio-technical configuration that should be upscaled and replicated in other Indian cities. But other actors are unsure about whether this actually facilitates the promised shift away from the private car and they lament how building this experimental system was predicated on excluding marginalized vendors and now-displaced slum-dwellers (Mahadevia *et al.* 2013).

In sum, we believe that what defines experiments in socio-technical transitions is their design as socially inclusive, practice-based and challenge-led projects and initiatives that aim at developing systemic innovation through a guided process of social learning that is robust in addressing ambiguity and uncertainty. Different aspects of this definition are highlighted across the transitions literature. Transition Management, for instance, explicitly foregrounds the challenge-led character of experiments, while Strategic Niche Management highlights ambiguity in values addressed in experimentation though the notion of second-order learning.

Challenges for urban governance: mobilizing the definition of experimentation

If cities are becoming a significant new arena for experiments for sustainability (Bai *et al.* 2010; Bulkeley *et al.* 2011, 2015; Romero Lankao 2012), then we anticipate a common set of governance challenges (Voss *et al.* 2006) to emerge as cities develop policies and arrangements to stimulate experiments. On the basis of our review of the literature on experiments, we have proposed a common definition of experiments. We believe that, in principle, all experiments can be assessed against this definition. Each of the dimensions we highlight points to a series of practical governance problems: how to ensure a requisite inclusivity of actors to participate in an experiment; how to calibrate the balance between public support and appropriability of knowledge in practice-based, open innovation processes; how to negotiate multiple values and interests in challenge-led projects; and how to design experiments so that they can be adaptive to uncertainty and unpredictability, while promoting valued and effective learning.

In terms of governance challenges, we believe there are three central governance problems, related to (1) coordination, (2) commitment and (3) negotiation.[7] First, in line with our ideas about unpredictability and ambiguity, there is the general coordination problem of the incentives and risks to participate in experiments amongst the diverse actors present in any given experimental context. We have argued that experiments are open and collaborative settings in which conventional incentives to invest in research, development and innovation through the generation of appropriable knowledge, technology and practice are likely to be weaker. Indeed, conventional models of innovation, focused on technological problem-solving within well-established techno-economic paradigms by business firms, do not describe the setting for learning and innovation that we observe in experiments. This suggests either that conventional incentives related to competitive advantage in established markets do not hold – although there may be additional incentives that are typical of conventional market settings, such as the generation of public goods, reputational benefits for firms or others, strategic positioning by firms that play a role – or that alternative ways of providing such incentives need to be built into the design of the experiment. Given the high risk of failure inherent in experiments, risk-sharing through collaboration, joint ventures and public–private partnerships will likely be common. One of the reasons why cities have become significant arenas for experimentation may be that they are appropriate venues to motivate and organize denser networks of collaboration amongst diverse actors who mutually benefit from information and other spillovers.

Second, there is the problem of mobilizing the required resources and long-term financial and political commitments needed to address a societal challenge, and to achieve system innovation through experimentation. In conventional innovation models, the development phase of R&D is in many sectors seen as being at least an order of magnitude more costly than research. Experiments intended to generate results and to contribute to regime transformation – and not instruments of symbolic politics (Lezaun 2011) – will require sustained and high levels of financial, social and political support. The problem of resourcing is a serious consideration at the urban level. These requisite resources, and with the stability and continuity needed, may not all be available at the urban scale. Sengers and Raven (2015) reveal that national governments are often crucial funders of urban experiments, while Wieczorek *et al.* (2015) argue that experiments may be supported by flows of resources across different scales, including transnationally. Nevertheless, the suitability of urban centres for experiments will depend on the availability of sufficient resources at the city level leveraged directly or indirectly. The existence of urban cultures fostering more sustainable lifestyles (McCormick *et al.* 2013), a specifically urban demand for innovative services, coupled with permissive local regulation and planning, also represent resources needed for experimentation to thrive.

Third, because there is ambiguity involved in experimenting for radically different socio-technical configurations, we have identified the problem of the negotiation of interests and values in the initial framing and through the course of experiments. This is a result of the broad actor base through which the experiment is constituted. Procedures for goal-setting, monitoring and evaluation and dispute resolution will be required throughout the life of experiments. As with the generation of resources for experiments, cities typically have more limited powers to develop regulatory, distributive and normative policies than nation states – although there are clearly diverse patterns of power at the city level, with some exercising more authority than others. In the novel, fluid and open process of learning within an experiment, this may be an advantage. Local urban contexts – where specific goals are set, where proximate social actors negotiate emerging obstacles and problems quickly and flexibly, and where procedures of feedback and accountability are developed – can provide advantages to experimentation. Close-knit communities of practice provide the potential for higher levels of trust and rapid decision-making. The capacity to put in place procedures to negotiate interests and values may be greatest at the urban (or neighbourhood) level.

More generally, modes and instruments of governance exist to address each of these challenges. Classical science, technology and innovation policy has been concerned with addressing the incentives/appropriability problem in innovation and there is a wide range of instruments, including tax incentives, subsidies, intellectual property rights, and standards and regulation, that have been used to encourage investments in innovation. In a recent review, the influential economist Dani Rodrik (2008) puts forward a series of new arguments for public intervention in (national) industrial and technological development. He argues that there are three features of good institutional design for industrial policy, which are relevant and fitting for our discussion of governance challenges in a similar context of experiments, prone to similar market and policy failures. These three are: embeddedness; 'carrots and sticks'; and accountability. By embeddedness, Rodrik alludes to a strategic collaboration and coordination between government and the private sector 'with the aim of uncovering where the most significant bottlenecks are, designing the most effective interventions, periodically evaluating the outcomes, and learning from the mistakes being made in the process' (Rodrik 2008: 27). By 'carrots and sticks', he refers to the need to combine generous fiscal and other incentives to encourage private-sector investment for innovation, while also setting aggressive performance targets which are then enforced. Finally, by accountability Rodrik refers to the need for risks and costs taken on by the public sector to incentivize participation to be transparent and for accountability to be clear. These principles are useful to address the incentive problem in the more open and collaborative settings of the experiment. In short, many of the issues we have identified in relation to governance of urban sustainability experiments are reflected in broader policy debates about the role of government policy in fostering innovation.

Conclusions: experimenting in the city

Based on our understanding of the transitions literature, we have defined an experiment as 'an inclusive, practice-based and challenge-led initiative designed to promote system innovation through social learning under conditions of uncertainty and ambiguity'. We have discussed each of these dimensions to point to a number of urban governance challenges and opportunities related to coordination of actors involved in experiments, sustaining resource commitments to radical innovations, and the negotiation of uncertainties and interests in experiments. It is clear that experimentation is a long-standing concept that started with a focus on creating niches for sustainable technologies and that, more recently, has broadened in scope and complexity to emphasize the distance between conventional notions of 'development' in science, technology and innovation policy, and the idea of an experiment in the context of sustainability transitions. This signals a bourgeoning field of practice on the ground and of research, which is likely to expand both conceptually and empirically in the future. Our definition of experiments and the discussion is intended to provide added focus to a rapidly developing debate. To conclude, we would like to point to a few promising avenues for further exploration in studying experiments as city-level entities.

First, we recognize that there is a difference between analysing experimentation as geared toward changing national socio-technical regimes and analysing experimentation geared to changing city-level regimes. Urban experiments are embedded in field-level structures (i.e. so-called 'global niches' not directly based on a spatial logic – see Geels and Raven 2006), as well as territorial structures (i.e. neighbourhoods, cities, regions or other institutions which are directly based on a spatial logic – see Hess 2004). Experiments are, after all, geared to transform regime structures, but the question remains how to best define these structures and on what kind of spatial logic they are based. Geels (2011: 14) argued that urban experiments should be seen in the context of changing national socio-technical regimes, allegedly 'because cultures, infrastructures, regulations and institutions are mostly (but not always) national phenomena'. Others emphasize that experiments challenge regime stability at the level of the city and focus on the obduracy of socio-technical systems inscribed onto the urban environment (Hommels 2005). Compared to the territories of nation states, cities are compact environments and infrastructural hubs where multiple socio-technical regimes are integrated within a confined space. Much work on sustainability transitions has gone into analysing the incumbency of national socio-technical regimes and the ways in which they might be changed through experimentation (Raven *et al.* 2012). Less attention has been paid to transforming highly interconnected and materially obdurate city-level socio-technical systems.

Important questions are thus related to how proximities in local and regional networks, infrastructures, resource endowments, political agendas, market structures, cultural settings and so on influence the form and outcome of experimental initiatives. And, vice versa, how these local spatial networks and processes can be reconfigured through experimentation. Such an agenda, when sensitive to the historical, path-dependent conditions in local or regional settings, would be able to improve our understanding of why experiments are more successful in some cities than in others (also see Coenen *et al.* 2010; Sengers and Raven 2015).

Finally, a promising avenue for further exploration involves the ways in which experiments connect across different spatial scales, perhaps through the facilitation of highly mobile change agents (Larner and Laurie 2010; McCann 2011). Research has demonstrated empirically that whilst local or national actors often initiate experimentation, transnational linkages are omnipresent (Sengers and Raven 2015; Wieczorek *et al.* 2015). The ways in which these multi-scalar structures influence experimentation positively or negatively, and the kinds of resources that flow through them, deserves further attention, including a better understanding of how experimental activity can diffuse across national borders and geographical scales.

Acknowledgements

Research supporting this chapter was funded by the Netherlands Organisation for Scientific Research NWO-WOTRO under the ‘Experimenting for Sustainability in India and Thailand’ project (grant number W 01.65.330.00) and by the European Commission FP7 Programme under the ‘Exploring Transitions Pathways to Sustainable, Low Carbon Societies (PATHWAYS)’ project (grant number 603942).

Notes

1 The field of ‘sustainability transitions’ is made up of a community of scholars who read and cite each other’s work on structural change toward green and inclusive socio-technical systems. For an overview of this emerging field, see Markard *et al.* 2012.

2 In Sengers *et al.* (2014), we describe the rationale of a ‘systematic review’ (Petticrew and Roberts 2006) and how we collected a dataset of 121 contributions (journal articles, book chapters, etc.) about experimentation from within the field of ‘sustainability transitions’. This dataset also serves as the basis for the second section of this chapter.

3 Another important term that has recently entered the transitions field is ‘living laboratory’ (Evans and Karvonen 2011) or ‘urban transition lab’ (Nevens *et al.* 2013). Because this does not point to a type of experiment *per se* but rather to experimentation as a mode of urban governance, we do not address it in this chapter.

4 The irony is that the most successful case of experimentation described with the SNM approach – in a way a foundational case study for the approach – was car sharing in

the 1990s (Hoogma *et al.* 2002). As part of developing the SNM approach, Harms and Truffer (1998) and Truffer (2003) describe how citizen cooperatives in Switzerland were crucial for the development of this type of 'social innovation'.

5 This is apparent when we look, for example, at the literature on sustainability experiments, some of which are described with the classic SNM approach reserved for niche experiments (Verbong *et al.* 2010). Other niche/transition experiments are described by combining the SNM and TM frameworks (Raven *et al.* 2010), while some of the literature on bounded socio-technical experiments mobilizes the same grassroots framing as grassroots experiments (Vergragt and Brown 2012).

6 The term 'social learning' (Bandura 1977; Argyris and Schön 1978) is explicitly used in the literature on bounded socio-technical experiments to highlight shifts in mindset (see Brown *et al.* 2003). Its counterpart in SNM is 'reflexive learning' or 'second-order learning' (Grin and van de Graaf 1996; Schot and Geels 2008).

7 Bulkeley *et al.* (2015) identify three similar factors influencing 'climate change experiments' in urban settings: institutional capacity in terms of remit, autonomy and capacity to act by municipal administrations; the availability of resources; and political will and leadership.

References

Argyris, C. and Schön, D. (1978). *Organizational Learning: A Theory of Action Perspective*. Reading: Addison Wesley.

Bai, X., Roberts, B. and Chen, J. (2010). Urban sustainability experiments in Asia: patterns and pathways. *Environmental Science and Policy*, 13: 312–325.

Bandura A. (1977). *Social Learning Theory*. Englewood Cliffs, NJ: Prentice-Hall.

Berkhout, F., Angel, D. and Wieczorek, A.J. (2009). Asian development and sustainable socio-technical regimes. *Technological Forecasting and Social Change*, 76: 218–228.

Berkhout, F., Verbong, G., Wieczorek, A.J., Raven, R.P.J.M., Lebel, L. and Bai, X. (2010). Sustainability experiments in Asia: innovations shaping alternative development pathways? *Environmental Science and Policy*, 13: 261–271.

Berkhout, F., Wieczorek, A.J. and Raven, R.P.J.M. (2011). Avoiding environmental convergence: a possible role for sustainability experiments in latecomer countries? *International Journal of Institutions and Economies*, 3: 367–385

Bos, J.J. and Brown, R.R. (2012). Governance experimentation and factors of success in socio-technical transitions in the urban water sector. *Technological Forecasting and Social Change*, 79: 1340–1353.

Brown H.S. and Vergragt, P.J. (2008). Bounded socio-technical experiments as agents of systemic change: the case of a zero-energy residential building. *Technological Forecasting and Social Change*, 75: 107–130.

Brown, H.S., Vergragt, P.J., Green, K. and Berchicci, L. (2003). Learning for sustainability transition through bounded socio-technical experiments in personal mobility. *Technology Analysis and Strategic Management*, 15: 291–315.

Bulkeley, H., Castán Broto, V., Hodson, M. and Marvin, S. (eds) (2011) *Cities and Low Carbon Transitions*. London: Routledge.

Bulkeley, H., Castán Broto, V. and Edwards, G.A.S. (2015). *An Urban Politics of Climate Change: Experimentation and the Governing of Socio-Technical Transitions*. London: Routledge.

Carvalho, L., Mingardo, G. and van Haaren, J. (2012). Green urban transport policies and cleantech innovations: evidence from Curitiba, Göteborg and Hamburg. *European Planning Studies*, 20: 375–396.

Castán Broto, V. and Bulkeley, H. (2013). A survey of urban climate change experiments in 100 cities. *Global Environmental Change*, 23: 92–102.

Coenen, L., Raven, R.P.J.M. and Verbong, G. (2010). Local niche experimentation in the energy transitions: a theoretical and empirical exploration of proximity and disadvantages. *Technology in Society*, 32: 295–302.

Cohen, W.M. (2010). Fifty years of empirical studies of innovative activity and performance. In Hall, B. and Rosenberg, N. (eds), *Handbook of the Economics of Innovation Vol.1*. Amsterdam: North-Holland, 129–213.

Dosi G. (1982). Technological paradigms and technological trajectories: a suggested interpretation of the determinants and directions of technical change. *Research Policy*, 11: 147–162.

Evans, J. and Karvonen, A. (2011). Living laboratories for sustainability: exploring the politics and epistemology of urban transitions. In Bulkeley, H., Castán Broto, V., Hodson, M. and Marvin, S. (eds), *Cities and Low Carbon Transitions*. London: Routledge, 126–141.

Geels, F.W. (2002). Technological transitions as evolutionary reconfiguration processes: a multi-level perspective case-study. *Research Policy*, 31: 1257–1274.

Geels, F.W. (2011). The role of the cities in technological transitions: analytical clarifications and historical examples. In: Bulkeley, H., Castán Broto, V., Hodson, M. and Marvin, S. (eds), *Cities and Low Carbon Transitions*. London: Routledge, 13–28.

Geels, F.W. and Raven, R.P.J.M. (2006). Non-linearity and expectations in niche-development trajectories: ups and downs in Dutch biogas development (1973–2003). *Technology Analysis and Strategic Management*, 18: 375–392.

Gibbons, M. (1999). Science's new social contract with society. *Nature*, 402: C81–84.

Gordon, I.R. and McCann, P. (2005). Innovation, agglomeration and regional development. *Journal of Economic Geography*, 5: 523–543.

Grin, J. and van de Graaf, H. (1996). Implementation as communicative action: an interpretive understanding of the interactions between policy makers and target groups. *Policy Sciences*, 29: 291–319.

Harms, S. and Truffer, B. (1998). *The Emergence and Professionalisation of Two Car-sharing Co-operatives in Switzerland: A Case Study for the Project 'Strategic Niche Management as a Tool for Transition to a Sustainable Transportation System'*. Dübendorf: EAWAG.

Hess, M. (2004). 'Spatial' relationships? Towards a reconceptualization of embeddedness. *Progress in Human Geography*, 28: 165–186.

Hoffman, M.J. (2011). *Climate Governance at the Crossroads: Experimenting with a Global Response after Kyoto*. Oxford: Oxford University Press.

Hommels, A. (2005). Studying obduracy in the city: towards a productive fusion between technology studies and urban studies. *Science, Technology, and Human Values*, 30: 323–351.

Hoogma, R., Kemp, R., Schot, J. and Truffer, B. (2002). *Experimenting for Sustainable Transport: The Approach of Strategic Niche Management*. London: Spon Press.

Jolly, S., Raven, R.P.J.M. and Romijn, H. (2012). Upscaling of business model experiments in off-grid PV solar energy in India. *Sustainability Science*, 7: 199–212.

Kemp, R., Schot, J.W. and Hoogma, R. (1998). Regime shifts to sustainability through processes of niche formation: the approach of strategic niche management, *Technology Analysis and Strategic Management*, 10: 175–196.

Kim, L. (1997). *From Imitation to Innovation: Dynamics of Korea's Technological Learning*. Boston, MA: Harvard Business School Press.

Larner, W. and Laurie, N. (2010). Travelling technocrats, embodied knowledges: globalising privatization in telecoms and water. *Geoforum*, 41: 218–226.

Lezaun, J. (2011). Offshore democracy: launch and landfall of a socio-technical experiment. *Economy and Society*, 40: 553–581.

Loorbach, D. (2007). *Transition Management: New Modes of Governance for Sustainable Development*. Utrecht: International Books.

McCann, E. (2011). Urban policy mobilities and global circuits of knowledge: toward a research agenda. *Annals of the Association of American Geographers*, 101: 107–130.

McCormick, K., Anderberg, S., Coenen, L. and Neij, L. (2013). Advancing sustainable urban transformation. *Journal of Cleaner Production*, 50: 1–11.

Mahadevia, D., Joshi, R. and Datey, A. (2013). Ahmedabad's BRT system: a sustainable urban transport panacea? *Economic and Political Weekly*, 48: 56–64.

Markard, J., Raven, R. and Truffer, B. (2012). Sustainability transitions: an emerging field of research and its prospects. *Research Policy*, 41: 955–967.

Marshall, A. (1920). *Principles of Economics*. London: Macmillan and Co.

Meadowcroft, J. (2011). Engaging with the politics of sustainability transitions. *Environmental Innovation and Societal Transitions*, 1: 70–75.

Nelson, R.R. and Winter, S.G. (1982). *An Evolutionary Theory of Economic Change*. Cambridge, MA: Belknap Press/Harvard University Press.

Nevens, F., Frantzeskaki, N., Loorbach, D. and Gorissen, L. (2013). Urban transition labs: co-creating transformative action for sustainable cities. *Journal of Cleaner Production*, 50: 111–122.

Oxford Dictionaries (2015). Definition of 'experiment' [Online]. Available: http://www.oxforddictionaries.com/definition/english/experiment?searchDictCode=all [1 February 2015].

Petticrew, M. and Roberts, H. (2006). *Systematic Reviews in the Social Sciences: A Practical Guide*. Oxford: Blackwell.

Raven, R.P.J.M., van den Bosch, S. and Weterings, R. (2010). Transitions and strategic niche management: towards a competence kit for practitioners. *International Journal for Technology Management*, 51: 57–74.

Raven, R.P.J.M., Schot, J. and Berkhout, F. (2012). Space and scale in socio-technical transitions. *Environmental Innovation and Societal Transitions*, 4: 62–78.

Rodrik, D. (2008). *Normalizing Industrial Policy: Commission on Growth and Development Working Paper 3*. Washington, DC: World Bank.

Romero-Lankao, P. (2012). Governing climate change in cities: an overview of policy and planning challenges and options. *European Planning Studies*, 20: 7–26.

Rotmans, J., Kemp, R. and van Asselt, M.B.A. (2001). More evolution than revolution: transition management in public policy. *Foresight*, 3: 15–32.

Schneidewind, U. (2012). Cities as 'real-world laboratories' for system innovations – theories, models and empirical Designs. Paper presented at *3rd International Conference on Sustainability Transitions*. Copenhagen, 29–31 August 2012.

Schot, J.W. and Geels, F.W. (2008). Strategic niche management and sustainable innovation journeys: theory, findings, research agenda and policy. *Technology Analysis and Strategic Management*, 20: 537–554.

Sengers, F. and Raven, R.P.J.M. (2014). Metering motorbike mobility: informal transport in transition? *Technology Analysis and Strategic Management*, 26: 453–468.

Sengers, F. and Raven, R.P.J.M. (2015). Toward a spatial perspective on niche development: the case of Bus Rapid Transit. *Environmental Innovation and Societal Transitions*, earlyview online.

Sengers, F., Wieczorek, A.J. and Raven, R.P.J.M. (2014). The role of experiments in sustainability transitions: a systematic literature review. Paper presented at *Studying Sustainability Transitions in Welfare States Workshop*. Rotterdam, 10–12 September 2014.

Seyfang, G. (2010). Community action for sustainable housing: building a low-carbon future. *Energy Policy*, 38: 7624–7633.

Seyfang, G., and Haxeltine, A. (2012). Growing grassroots innovations: exploring the role of community-based initiatives in governing sustainable energy transitions. *Environment and Planning C*, 30: 381–400.

Seyfang, G. and Smith A. (2007). Grassroots innovations for sustainable development: towards a new research and policy agenda. *Environmental Politics*, 16: 584–603.

Smith, A. and Raven, R.P.J.M. (2012). What is protective space? Reconsidering niches in transitions to sustainability. *Research Policy*, 41: 1025–1036.

Truffer, B. (2003). User-led innovation processes: the development of professional car sharing by environmentally concerned citizens. *Innovation*, 16: 139–154.

van de Poel, I. (2000). On the role of outsiders in technological development. *Technology Analysis and Strategic Management*, 12: 383–397.

van den Bergh, J.C.M. (2012). EIST one year: something to celebrate? *Environmental Innovation and Societal Transitions*, 4: 1–6.

van den Bosch, S. (2010). *Transition Experiments: Exploring Societal Changes Towards Sustainability*. PhD Dissertation, University of Rotterdam.

van den Bosch, S. and Rotmans, J. (2008). *Deepening, Broadening and Scaling up: A Framework for Steering Transition Experiments*. Rotterdam: Knowledge Centre for Sustainable System Innovations and Transitions.

Verbong, G.P.J., Christiaens, W., Raven, R.P.J.M. and Balkema, A. (2010). Strategic niche management in an unstable regime: biomass gasification in India. *Environmental Science and Policy*, 13: 272–281.

Vergragt, P.J. and Brown, H.S. (2012). The challenge of energy retrofitting the residential housing stock: grassroots innovations and socio-technical system change in Worcester, MA. *Technology Analysis and Strategic Management*, 24: 407–420.

Verheul, H. and Vergragt, P. (1995). Social experiments in the development of environmental technology: a bottom-up perspective. *Technology Analysis and Strategic Management*, 7: 315–326.

Voss, J.P., Bauknecht, D. and Kemp, R. (2006). *Reflexive Governance for Sustainable Development*. Cheltenham: Edward Elgar.

Wieczorek, A.J., Raven, R.P.J.M. and Berkhout, F. (2015). Transnational linkages in sustainability experiments: the case of solar PV in India. *Environmental Innovation and Societal Transitions*, earlyview online.

3 Cities, experiments and the logics of the knowledge economy

Tim May and Beth Perry

Introduction

> The city is a completely interacting system and thus, the experiment must be a total system. Nobody knows the answers to city living in the future, and, when answers are unknown, experiment is essential.
>
> (Spilhaus 1967: 1141)

Such were the concluding remarks of Athelstan Spilhaus' 1967 essay on 'The Experimental City', written in a special edition of *Daedalus* dedicated to America's changing environment. A scientist, administrator and educator, Spilhaus then worked with local state officials to develop an ambitious plan for the 'Minnesota Experimental City' during the 1970s, a proposed 'test-tube' city, intended to act as a laboratory to solve the twin ecological and social crises of the times. Underpinning his vision was an experimental logic grounded in the central idea that city-scale solutions are needed to develop the city as a 'completely interacting system'. This chapter takes this city-scale seriously by examining how global dynamics and the pervasive logics of the knowledge economy frame the contexts and conditions within which experimentation takes place. We contend that the ideas and ideals of urban experimentation cannot be isolated or analysed separately from the dynamics of neoliberal capitalism, if their motivations, possibilities and limits are to be more rigorously understood. Cities, after all, are positioned as critical places where social, political, economic and ecological challenges in the twenty-first century will be met. In these processes, knowledge is accorded a central role in aspirations for economic growth, competitiveness and sustainability. Political desires and formal policy frameworks at international, national and sub-national levels have positioned cities, in different ways, to play a key role in harnessing science, technology and innovation towards these ends.

City elites and urban growth coalitions share common aspirations reflected in the search for symbolic and material advantage, which includes being attractive to inward investment by global capital. The promise of the knowledge economy is reflected in rhetorical flourishes accompanied by geological imagery: corridors, clusters, capitals and valleys to alleys, glens and fens. Expectations in these contexts are high; knowledge is attributed with the power to re-make places, accompanied by significant improvement in economic, social, cultural, educational, health and environmental outcomes. Cities seek these outcomes with recipes for growth whose origins and perpetuation lie not just in the practices of consultancies but also in academia (O'Mara 2005; Perry and May 2011). Aspiring cities frequently see their universities as strategic actors towards common goals, which, by virtue of their estates or specific engagement activities and similar aspirations to be global, become part of the aspiring mix (May and Perry 2006). At a city-scale, the logics of the knowledge economy tend to ignore context in search of symbolic and resource competition with other cities.

The idea of an experimental city can be viewed against this backdrop. Whilst urban scholars have examined the sustainable city (for example, see Flint and Raco 2011), less emphasis is placed on the relationships between knowledge, the city and experimentation from a social epistemological point of view in the climate of neoliberal capitalism. In other words, there is little research on how 'knowledge' is implicated more specifically within urban strategies and how 'experimentation' is attributed with value in this political climate. In the process, important issues are raised not only on the relationship between theory, policy and practice, but also the intended and actual beneficiaries of urban knowledge-based development, as well as the future of the university as a site of critical knowledge production (May and Perry 2011a). Our contribution is to examine a particular logic of experimentation, rather than a single case, a logic that is framed by the discourses and practices of the knowledge economy. Our argument draws on widespread literature reviews and secondary sources and work undertaken over 15 years in comparative research programmes with cities in the UK, Europe and Africa, including interviews, workshops and observations. Elsewhere, we have written about how global dynamics are mediated and embedded in different ways in these cities. Here, we want to make visible the underlying dynamics and discourses that saturate a number of city positions and strategies (see also May and Perry 2016).

The chapter starts by examining the relationships between cities and their desire for economic growth – through the lens of the knowledge economy – as a context within which more localised innovations sit. We question how experimentation fits within the search for cities to be innovative, creative and smart, in light of the increasing critique of these urban 'technopoles of the world' (Miao *et al.* 2015). Then, we examine the idea of experimentation in terms of the forms taken by cities in contemporary capitalism and the attribution by politicians, officials and expert cultures to its potential. Our focus is not the experiment itself, but rather *experimentalism* as an orientation that favours experimentation and innovation. The chapter constitutes a provocation, which we argue, should give cause for circumspection and modesty in stating claims for the grand experimental city. We suggest that experimentation may risk denying the importance of context in ignoring history and place-based dynamics and that there is a fundamental mismatch between experimentation and the ontology and epistemology of the city. We end with a note of caution – to ensure that the promise of experimentation does not become captured by the usual urban elites, as nothing more than a side-stepping of the deep structural inequalities facing urban societies. Indeed, like Ong (2011: 5) we too are struck by the 'paradoxical interdependence of calculative practices of political entrepreneurialism and the progressive language of anti-neoliberalism' as manifest in the differential deployments of the language and practice of 'experimentation' in the contemporary city.

Cities and the knowledge economy

The knowledge-based economy has been regarded as a development equal in importance to the Industrial Revolution of the nineteenth century. Whilst knowledge has always played an important role in human activities (Stehr 1994), its quantity, complexity and speed marks a new economic phase (de Weert 1999: 52). This premise is based on the recognition that knowledge is increasingly a source of competitive advantage leading to the search for 'new ways of producing, using and combining diverse knowledges; the same ingredients ... rearranged in new and better recipes' (Bryson *et al.* 2000: 1). Building 'high value added' economies, characterised by high wages, employment and skills is increasingly seen as a priority. The increasing volume of workers involved in distributing, processing and producing knowledge, along with the percentage of GNP (gross national product) and salaries to specific business sectors have all been taken to signify the coming of what has been termed the 'information age' (Bell 1979). The causes of these changes have been attributed to globalisation, the proliferation of high-tech industries, expansion of the scientific base, movement from manufacturing to a service-based economy, new information technologies and accelerated technological changes (Neef 1998).

The development of the knowledge economy requires a re-evaluation, reconfiguration and reconstitution of ideas and actions as a precondition for improved understanding at different spatial scales (Scott 2000). Much has been written about the 'death of distance' and the 'end of geography' (Morgan 2001) and the increasing globalisation and internationalisation of markets, economies, societies and environments has been taken to preface the breakdown of the boundaries of time and space. However, far from signalling the end of face-to-face and 'human' interactions, globalisation has also been characterised as requiring an increasing emphasis on regions, localities and cities as engines of economic growth (Savitch and Kantor 2004). Knowledge-based wealth creation requires economies of scale, clusters and a critical mass of complementary expertise within a particular location such that spillover then leads to innovation and productivity (Florida 2002). Firms draw on location-specific factors to ensure competitive success and from there upon the resources within local environments (Simmie *et al.* 2002).

This discourse is reflected in national and urban initiatives in which science and innovation have been placed at the heart of economic growth and competitiveness. A dominant view of the relationship between knowledge and cities in policy literatures emphasises an instrumentally driven, econocentric perspective on cities and their contributions to national wealth and competitiveness. Its language includes 'drivers', 'opportunities', 'pillars', 'initiatives' and 'solutions' (see, for example, Dresner, 2001). In the UK during the 2000s, this was represented in the Science Cities initiative (May and Perry 2011a; Perry and May 2015), whilst elsewhere in innovative business districts, clusters of excellence and grand technological platforms. Commonly, cities have begun to emphasise physical site development, regeneration and infrastructure through, for example, a focus on hubs, incubators and science parks (Youtie and Shapira 2008). Many developments centre around technology as a key driver and enabler, creating 'Invented Edens'. As Kargon and Mollela (2008: 12) argue, 'the techno-city represents an experiment in integrating modern technology into the world of ideal life'.

On first glance, place, it seems, still matters economically (Sassen 2012). In pursuing strategies of mitigation and adaptation in the face of the global carbon crisis, meso and micro dynamics are seen to be significant factors (Busch and Shrivastava 2011). A number of case studies have emerged to show how different cities approach the challenges of knowledge-based growth, from Manchester and Eindhoven, to Barcelona, to Holon and Singapore (Wong *et al.* 2006; Clua and Albet 2008; Ooi 2008; Fernandez-Maldono and Romein 2010; May and Perry 2011a). Emphasis is placed on different pathways to development, success factors, historical trajectories and the consequences and limitations of such approaches. Yet despite this, cities that aspire to be global in their aspirations want global universities and together they form part of the clamour for symbolic advantage (see May and Perry 2006; Perry and May 2011). An increased relevance attached to knowledge has led to a plethora of actors focusing upon and positioning themselves within a sphere of activity characterised by what seems to be limitless potential. Efforts to reshape regional and local identities through harnessing the 'brand' power of science and technology have been charted (Brenner 2004; Perry and May 2007). So we have knowledge capitals, silicon alleys, BioValleys, digital and smart cities as the apparent foundation for strategic, competitive success, in a desire to create exclusive environments (O'Mara 2005).

In this context, the 'Experimental City' becomes little more than the new sub-brand, a world-class accolade sought to differentiate and position cities in global knowledge hierarchies. Greater Lyon, France, is a case in point, which proudly proclaims itself to be a 'Smart City – Experimenting Today for Better City Living Tomorrow' (Only Lyon 2014). Whilst embracing ecological and technological innovation, such a strategy is for the few not the many, led by business interests and clearly positioned to attract international capital investment. When it comes to understanding these dynamics, we cannot simply separate the social from the environmental, nor allude to these changes as if a reflection of some economic natural order. They are the product of power relations operating at an international level whose intellectual underpinnings, on the occasions they are publicly contested and called upon to justify themselves, represent a new doxa:

> Like the Catholic *doxa* of the Middle Ages, the new neoliberal *pensée unique* seemed to provide solutions for all kinds of social and ecological issues. Often neutralised in academic terms and amplified by associated intellectuals within and outside the mainstream media, these solutions are relentlessly preached – not least to students who will constitute the elites of the future.
>
> (Koch 2012: 190)

The evaporation of context

Post-2008, enthusiasm for the city as a site of speculative accumulation has increased (Harvey 2012). Conspicuous consumption, the power of multinational corporations and ever-increasing inequality all play their role in a system that was on the brink of collapse until states and their citizens provided welfare payments for the consequences of the actions of those in the financial sector (Gamble 2009; Crouch 2011; Dorling 2014). The push for ever greater centres of excellence, networked across the globe, is accelerating, along with a preference for forms of knowledge and their representation (Chapain *et al.* 2009; May and Perry 2011b). What is illuminated is the regressive nature of policy with its narrow economic focus provided through the exclusion of cultural factors (Scott 2014). Opportunities are 'colonized' (Wyly 2013) within a process of capital accumulation, whose effects relate directly to how knowledge is seen, deployed and interpreted (Lave 2012). It is within this mix that the idea of the experimental city has gained ascendancy. Politics, conducted in the imaginary of globalisation, forms a wide gap between actuality and potentiality: 'in such globalising circumstances, the neoliberal as a global form comes to articulate stated experimentations with an art of being global' (Roy and Ong 2011: 3). Existing reality is the wicked problem of the present, whilst there is no apprehension in the future but varying forms of comprehension in anticipation of its transcendence.

At the city-scale, 'success' is not judged on community-based alternatives – rather, according to the aspirations of officials, politicians and the 'experts' who are the intellectual jugglers of the permanent possibilities that feed neoliberal ideologies. These are the actions of those tinkering 'with machines and models to demonstrate the mechanics of economic interaction' (Gibson-Graham *et al.* 2013: 1). This all takes place on the terrain of the Imaginary: markets, after all, are about belief in belief (Žižek 2009). More often than not, cities are positioned differentially, with some seen to be 'lagging behind' (Pike *et al.* 2006) and subsequently assumed to be characterised by second-rate thinking and practice. No wonder that urban politicians, corporate managers and officials are not able to publicly admit of difficulties and instead speak only of 'moving forward'. This produces a falsity: localism is assumed to be able to harness the potentiality in global neoliberalism to advance the prospects of cities; yet global neoliberalism relies upon the very inequalities that have produced uneven spatial development in the first place! In this climate, governments frame their roles as creating 'opportunities', whilst actively or inadvertently reinforcing the conditions for this state of affairs to flourish. Places judge themselves and others according to how imaginatively they have responded to opportunities provided in the neoliberal economy. Anything less than full embrace of the promise of opportunities in an imaginary future is seen as a challenge and/or the reactivation of an apparently outdated politics of need. Symptomatic politics is allowed a free reign and holds up privileged places as 'leading the game' in the name of improving other cities. In the competitive politics of global-urban hierarchies, cities embrace experimentalism to demonstrate to other cities, selected entrepreneurs, innovators and universities that they mean business. In the process, particular cities are heralded as emblematic of how things can be achieved and thus serve as models for others to emulate. It is as if governmental support is not important, as they become nothing more than the logical locations for aspirations – leaving intact the myth of global capital as inherently mobile and indiscriminate. References to growth, opportunity and the transactions around emblematic status for others to seek to emulate, without sensitivity to context and capability, are then permitted an uninterrupted path to the benefit of elites.

The idea of experimentation is part of the story through which knowledge has not only been produced, but also validated. The idea of the context of experimentation is that it needs to be controlled to avoid undermining the desire for knowledge. Orderly investigation of natural phenomena rules out those factors such as political belief on the grounds not only of validity, but replicability: that is, the production of truth as a public act can take place through different individuals witnessing the same experiment and coming to the same conclusions. In this way, 'collective witnessing would convert self-serving disputes into mutually accepted knowledge' (Poovey 1998: 113). The experiment may take place in a context, but it relies on being replicable, regardless of that context. How does this then fit with place-based dynamics? History evaporates, as the textures of communities are lost in the search for symbolic futures. Symbolism and global posturing appear to triumph over any sense of connection with the realities of urban life. For Lefebvre (1996), the city exhibits a double morphology, or form. It manifests itself in a social dimension: that is, how people communicate, move between places, engage in various activities and for various reasons and with a range of consequences. It also exhibits a material dimension in terms of building and spaces, some of which are inhabited and others that are uninhabited: for example, public buildings, monuments, squares and streets. He used the idea of implosion–explosion to capture how cities were both being destroyed but also growing to support industrialisation (Brenner 2014).

These context-sensitive issues may be readily dealt with in understanding how the idea of experimentation plays its role in urban economic growth aspirations. In comparing experimentation in the eco-cities of Shanghai and Tianjin, Miao and Lang (2014: 8) stress the strong logic of economic growth and argue that 'the chances for success become slimmer if experimental initiatives pursue more comprehensive goals, like providing social goods'. Experiments work, as they have done in their history, not in terms of testing hypotheses, but as demonstrators (see Poovey 1998; Pickstone 2000). They are about context-revision, not sensitivity. They demonstrate adaptability and flexibility in the face of aspiration under conditions of heightened uncertainty. The particular elements brought together within a city demonstrate global aspiration that signals attraction and flexibility for capital accumulation. In accounting for the lack of success attributed to the Shanghai 'Dongtan' project, Miao and Lang demonstrate its inability to act as a model for national policy-making or indeed to attract a powerful transnational investor. Here, we find apparent legitimacy for the adoption of a universal method that readily permits scientism to saturate the realm of political responsibility but also indicates willingness for the city to be a place of continual adaptation to the demands of neoliberal capitalism. The results demonstrate the receptiveness of the aspirant city to become a vibrant site for the future. The imagery of the method mixes with desire and is placed beyond doubt as cities become ever more flexible in their attraction to capital. It is a frenetic process without end: the intangible in search of the unattainable (May and Perry 2006).

The experimental method

Experimentation is bolstered by the desire for closure, control and certainty that characterises political decision-making, exacerbated by the intractability and complexity said to characterise 'wicked urban problems' (Polk 2014). This stands in contrast to the exercise of doubt more characteristic of academic discourse. Seeking the means to achieve democratic legitimacy means the need to convince an urban population of the efficacy of processes, whilst also knowing that their outcomes are not within control and may even lead to greater uncertainty. Thus:

> The political discourse of social change must make its appeal as wish-fulfilling dreams of omnipotent intentions or destined futures. From this perspective, every political programme is not only inherently fatalistic, but also mired in fantasy and narcissism from its inception. The discourse of social change takes place on the terrain of the Imaginary.
>
> (Rothenberg 2010: 153)

Politics begins with the denunciation of 'a tacit contract of adherence to the established order which defines the original doxa' (Bourdieu 1992: 127). Political action is informed by knowledge of the world and a capability to act on that knowledge. Whilst invoking experimentation may seem to challenge the existing states of affairs, it provides relief from ambivalence through a selectivity that omits so much (and so many) from its process. Whilst localised experiments may claim to be opening up technological change to plural interests, the symbolism of experimentalism at a city-scale appears to represent the ceaseless, careful empiricism of the scientific attitude whose practices constitute rules and procedures for practitioners across continents and time. It is 'ignorant' to be against that which has not yet found its results, for to do so is to stand in the way of progress. Criticism of the validity of its adoption questions the very pre-suppositions upon which the decision-making process is based. Questions of 'why' and 'for whom' become luxuries in face of those who see experimentation as part of an insatiable desire for economic growth.

What of the methodological flaws of experimentation when applied to the urban environment in all of its manifestations? For instance, the establishment of relations of cause and effect require that 'standing conditions' are specified to permit genuine comparison. That means introducing constants that presuppose closed systems. The city is comprised of open systems whose amenability to control is highly problematic. This raises particular problems when it comes to replication, as time, place, context and character cannot be reproduced in the same manner in different settings (May 2011). The idea of experimentation has a laudable aim, but the symbolism of scientism has been appropriated by political aspiration that tends to not only reproduce, but accelerate, a status quo riddled with contradictions (Harvey 2014). One consequence is the eradication, or highly selective interpretation, of history, but also the terrain of the 'political' itself. So, let us go back to ideas formed in the seventeenth century to understand what this does to the practice of contemporary urban politics and how that informs not the affirmation of present trajectories through supposed innovations for a minority, but transformations of our cities for the majority.

Giambattista Vico (1668–1744) was critical of those who take particular methods and insights and over-extend them into domains in which they are inapplicable. He was certainly not against science, but scientism. Disciplinary specialism was of importance, but when specialism was applied to areas that were foreign to understanding this was due to the 'weakness of our nature, which prompts us to take inordinate delight in ourselves and in our own pursuits' (Vico 1990: 80). Excessive attention to the potential of experimentation in the current political climate works to relieve politicians and experts of the responsibility for processes that should be part of public, political discourse, not disguised as technical-instrumental modes of reasoning. As Vico put it:

> We neglect that discipline which deals with the differential features of the virtues and vices, with good and bad behaviour-patterns, with the typical characteristics of the various ages of man, of the two sexes, of social and economic class, race and nation, and with the art of seemly conduct in life, the most difficult of all arts. As a consequence of this neglect, a noble and important branch of studies, i.e., the science of politics, lies almost abandoned and untended.
>
> (Vico 1990: 33)

It is not just the tension between facts and values that is being disguised in alignments amongst consultancies, academics and politicians here. The gap between the imaginary and the real is one between the passive scientific gaze and the active post-Kantian self in which ideas cannot and should not be squashed but explored and interrogated in confrontation with the realities of the world itself (May and Perry 2011c). In the history of science and experimentation, the overall result became 'an opposition between allegedly passive observation and active experimentation and a split within the scientist's own self' (Daston and Galison 2007: 242). We can see this same process at work in the co-optation of the idea of experimentation as it enables the future to be placed in a process that seeks to reconfigure the present in the name of the imaginary. The result is that a politics of the present is suspended in the name of a possibility that benefits the few.

When the above processes are aligned with the tendency to move towards more impersonal forms of knowledge in the relations between science and the public imagination (Ezrahi 2006), we can see the idea of experimentation as an ingredient in the reconfiguration of the relationship between knowledge and power in neoliberal development. There are evidently issues here concerning expertise and democracy in which the word 'quasi' seems apposite: 'quasi' experimentation and expertise. Political decision-making is often justified by allusion to scientific knowledge but no political decision can be justified through such means. For that to be the case, it would be a blatant conflation and we do not have to travel back too far in history to see the violent consequences of such a move. Also, as we have noted in our methodological asides, we are not dealing with closed systems, but open ones, and so the strict definition of scientific experimentation is not applicable. As we have said, it is attribution we are dealing with in the desire to attract capital to cities in the search for profit in a place. It does that by placing it in the realm of the factual, which is actually the domain of necessity through the denial of interpretation and choice, so the city can play in the global game of accumulation.

Alternative practices, particularly those that emanate from urban communities that are viewed as part of the economic wasteland as by-products of market forces, are sidelined. These forms of collaborative support tend not to be represented in favour of a particular set of descriptions and a focus upon current local deficits and the attraction and retention of particular groups who offer the seeds of a better future. The overall effect is a 'depoliticized simulation of truth' (Poster 1990: 62) which fails to recognise that 'the chances of translating knowledge for action into knowledge in action are immeasurably improved once it is recognized that the probability to realize knowledge is dependent on context-specific social, political and economic conditions' (Stehr 1992: 121).

Whilst administrations focus their efforts upon creating terrains of possibility through experimentation, a parallel process of disengagement with the realities of the present, including the causes of inequality, is assisted by representation through particular indicators, targets and metrics. These include measures which emphasise the ecological as if it were separate from social sustainability through a focus on environmental or social justice indicators rather than quality of life (Agyeman 2013). Politically and epistemologically, technocracy and empiricism are triumphant. This separation permits a spectator view of the urban that is characteristic of the passive element of the methodical scientist. When justification is called for in this frenetic pursuit, it may be found amongst those academics only too happy to provide it through the provision of their latest blueprints for success. Even amongst those who claim to be critical, they are subject to these processes and when it comes to reflexive and critical thinking, it is often absent (May 2005; May and Perry 2013). The need for partnerships in critique among those normally omitted from academia is particular important given that experimentation is part of the 'new spirit of capitalism' (Boltanski and Chiapello 2005), which can readily absorb critique and escape to other places (Chiapello 2013).

For a responsible urban experimentalism

When it comes to the attempt to represent and invoke the 'Real' who can possibly be 'against' experimentation? As Ann Oakley puts it,

> reality exists, and, although we do not all see it in the same way, we share an interest in being able to live our lives as well as we can, free from ill-informed intervention and in the best possible knowledge we can gather of what is likely to make all of us healthy, most productive, most happy and most able to contribute to the common good.
>
> (Oakley 2000: 323)

In addition, curiosity informs experimentation and offers the advantage of what is not yet known but may be realised if possibility is permitted its free reign. It has, after all, informed utopian thought which sees in the distance from the present a future of hope (Levitas 2011). In our times of change and flux, old ways of thinking are framed as rigid and new ways are required. Adaptation, flexibility and openness are seen as necessary predispositions in our 'risk societies' (Beck 1992). One could argue that urban experimentalism can be seen as part of this process – a way of controlling risk, in risk-averse public sector organisations, of bounding innovation in safe spaces, outside the usual pressures of urban policy-making. Heilmann (2008) argues that in China decentralised experimentation helps cushion the unknown impacts of political uncertainty and policy ambiguity. Urban experiments are therefore said to 'break the established horizons of urban standards in and beyond the city … they are experiments in that they put forth questions, initiatives and procedures in the midst of uncertainty, without guarantees about successful outcomes' (Roy and Ong 2011: 3).

In this chapter, we have sought to draw attention to an alternative narrative and framing. There is a danger that experimentation under conditions of neoliberal practices does not positively benefit the populations of a city but affirms an acceleration of a system that has produced staggering inequalities. Experimentalism can be seen as a way of testing responses to urban challenges without needing to attend to the structural inequalities or crises that may have given rise to those challenges in the first place. If experimentalism is a strategy for urban transformation, it is an 'affirmative' (Fraser 2003) one which runs the risk of bolstering business-as-usual outside the experimental bubble, tinkering at the edges of a broken system. What is often called a problem of poverty from very particular points of view is a problem of the riches of others (Sayer 2015). We are not alone in sounding a trumpet for a cautious rather than wholesale embrace of urban experimentalism. Oosterlynck and Gonzalez (2013: 1076) argue the 1970s economic crisis opened up the city as a laboratory for neoliberal experiments and that now, 'as we start to see a variety of urban responses unfold across the world, we must ask whether they serve to reinforce ongoing neoliberal urban restructuring or effectively produce new, post-neoliberal, urban governance rationalities.' It is for these and other reasons, that there are calls to end the experiment of the last 30 years with a return to the foundational economy (Bowman *et al.* 2014).

At the city-scale, the closing down or bracketing of difficult and intractable urban issues through the deployment of the 'experiment' has a different logic, we argue, from the logics manifest in grassroots initiatives. Grassroots experimentalism within cities can be seen as a response to the structural obduracy of the urban system, its interconnectedness with global flows of capital accumulation, a necessary bounding of systems to prove that there are alternative ways of living and working. In such a case, is the language of experimentalism deployed consciously to legitimise dissent that might otherwise be relegated as social protest? At the city-scale, a responsible urban experimentalism would take these grassroots initiatives seriously in terms of learning about potential alternatives to the dominant hegemony (see McFarlane 2011). Commitment to learning and transformation, coupled with the urban capacities and capabilities to act, are needed at the urban level. This, of course, begs the questions: a capacity to know what and a capability to do what? John Dewey (1957: 97) characterised rationalism as tending towards a 'carelessness, conceit, irresponsibility, and rigidity'. He also referred to 'experimental intelligence' which can liberate and enable action to be more directed and less blinkered and free us from 'the bondage of the past, due to ignorance and accident hardened into custom' (Dewey 1957: 96). This idea of knowledge and its component of experimentation are very different from the ones we have characterised. It is informed by a 'cooperative search for truth for the purpose of coping with real problems encountered in the course of action' (Joas 1993: 19). Dewey's emphasis upon the need for a pre-political basis of social cooperation introduces a corrective to the current one-sided politics where experts are mobilised to justify practices that benefit the few. Participation in public deliberations depends on having already experienced a communicative relatedness (Honneth 2007). Experimentation can become different if opened up to those from whom it is normally closed off, but its effects will be felt differently until such time that more equal and sustainable cities are achieved.

Acknowledgements

Thanks to the editors and other contributors for the comments on an earlier draft of this chapter. We also acknowledge the following funders of research underpinning this chapter: Mistra Urban Futures, Arts and Humanities Research Council (grant number AH/J005320/1) and ESRC (RES-151-25-0037).

References

Agyeman, J. (2013). *Introducing Just Sustainabilities: Policy, Planning, and Practice*. London: Zed Books.

Beck, U. (1992). *Risk Society: Towards a New Modernity*. London: Sage.

Bell, D. (1979). The social framework of the information society. In Dertouzos, M. and Moses, J. (eds), *The Computer Age: A 20 Year View*. Cambridge, MA: MIT Press, 500–549.

Boltanski, L. and Chiapello, E. (2005). *The New Spirit of Capitalism*. London: Verso.

Bourdieu, P. (1992). *Language and Symbolic Power*. Cambridge: Polity.

Bowman, A., Ertürk, I., Froud, J., Johal, S., Law, J., Leaver, A., Moran, M. and Williams, K. (2014). *The End of the Experiment? From Competition to the Foundational Economy*. Manchester: Manchester University Press.

Brenner, N. (2004). *New State Spaces: Urban Governance and the Rescaling of Statehood*. Oxford: Oxford University Press.

Brenner, N. (ed.) (2014). *Implosions/Explosions: Towards and Study of Planetary Urbanization*. Berlin: Jovis Verlag.

Bryson J., Daniels P., Henry N. and Pollard J. (eds) (2000). *Knowledge, Space, Economy*. London: Routledge.

Busch, T. and Shrivastava, P. (2011). *The Global Carbon Crisis: Emerging Carbon Constraints and Strategic Management Options*. Sheffield: Greenleaf.

Chapain, C., Collinge, C., Lee, P. and Musterd, S. (eds) (2009). Can we plan the creative knowledge city? *Built Environment,* 32 (Special edition).

Chiapello, E. (2013). Capitalism and its criticisms. In du Gay, P. and Morgan, G. (eds), *New Spirits of Capitalism: Crises, Justifications, and Dynamics*. Oxford: Oxford University Press, 60–81.

Clua, A. and Albet, A. (2008) 22@bcn: Bringing Barcelona forward in the information era. In Yigitcanlar, T., Velibeyoglu, K. and Baum, S. (eds), *Knowledge-Based Urban Development: Planning and Applications in the Information Era*. New York: Information Science Reference, 132–148.

Crouch, C. (2011). *The Strange Non-Death of Neoliberalism*. Cambridge: Polity.

Daston, L. and Galison, P. (2007). *Objectivity*. New York: Zone Books.

De Weert, E. (1999). Contours of the emergent knowledge society: theoretical debate and implications for higher education research. *Higher Education,* 38: 49–69.

Dewey, J. (1957). *Reconstruction in Philosophy*. Boston. MA: Beacon Press.

Dorling, D. (2014). *Inequality and the 1%*. London: Verso.

Dresner, S. (2001). A comparison of RTD structures in EU member states. In Dresner, S. and Gilbert, N. (eds), *The Dynamics of European Science and Technology Policies*. Aldershot: Ashgate, 109–135.

Ezrahi, Y. (2006). Science and the political imagination in contemporary democracies. In Jasanoff, S. (ed.), *States of Knowledge: The Co-Production of Science and Social Order*. London: Routledge, 254–273.

Fernandez-Maldono, A. and Romein, A. (2010). The role of organisational capacity and knowledge-based development: the reinvention of Eindhoven. *International Journal of Knowledge-Based Development,* 1: 79–97.

Flint, J. and Raco, M. (2011). *The Future of Sustainable Cities: Critical Reflections*. Bristol: Policy Press.

Florida, R. (2002). *The Rise of the Creative Class and How It's Transforming Work, Leisure, Community and Everyday Life*. New York: Basic Books.

Fraser, N. (2003). II. Integrating redistribution and recognition: problems in moral philosophy. In Fraser, N. and Honneth, A (eds), *Redistribution or Recognition: A Political-Philosophical Exchange*. London: Verso, 26–47.

Gamble, A. (2009). *The Spectre at the Feast: Capitalist Crisis and the Politics of Recession*. London: Palgrave Macmillan.

Gibson-Graham, J.K., Cameron, J. and Healy, S. (2013). *Take Back the Economy: An Ethical Guide for Transforming our Communities.* Minneapolis: University of Minnesota Press.

Harvey, D. (2012). *Rebel Cities: From the Right to the City to the Urban Revolution.* London: Verso.

Harvey, D. (2014). *Seventeen Contradictions and the End of Capitalism.* London: Profile.

Heilmann, S. (2008). Policy experimentation in China's economic rise. *Studies of Comparative International Development,* 43: 1–26.

Honneth, A. (2007). *Disrespect: The Normative Foundations of Critical Theory.* Cambridge: Polity.

Joas, H. (1993). *Pragmatism and Social Theory.* Chicago: University of Chicago Press.

Kargon, R. and Mollela, A. (2008). *Invented Edens: Techno-Cities of the 20th Century.* Cambridge, MA: MIT Press.

Koch, M. (2012). *Capitalism and Climate Change: Theoretical Discussion, Historical Development and Policy Responses*. Basingstoke: Palgrave Macmillan.

Lave, R. (2012). Neoliberalism and the production of environmental knowledge. *Environment and Society: Advances in Research,* 3: 19–38.

Lefebvre, H. (1996). *Writings on Cities.* Oxford: Blackwell.

Levitas, R. (2011). *The Concept of Utopia.* Oxford: Peter Lang.

McFarlane, C. (2011). *Learning the City: Knowledge and Translocal Assemblage*. Oxford: Wiley-Blackwell.

May, T. (2005). Transformations in academic production: context, content and consequences. *European Journal of Social Theory*, 8: 193–209.

May, T. (2011). *Social Research: Issues, Methods and Process, 4th Edition.* Maidenhead: McGraw-Hill.

May, T. and Perry, B. (2006). Cities, knowledge and universities: transformations in the image of the intangible. *Social Epistemology*, 20: 259–282.

May, T. and Perry, B. (2011a). Contours and conflicts in scale: science, knowledge and urban development. *Local Economy*, 26: 715–720.

May, T. and Perry, B. (2011b). *Social Research and Reflexivity: Content, Consequences and Context.* London: Sage.

May, T. and Perry, B. (2011c). Urban research in the knowledge economy: content, context and outlook. *Built Environment,* 37: 352–367.

May, T. and Perry, B. (2013). Universities, reflexivity and critique: uneasy parallels in practice. *Policy Futures in Education*, 11: 505–514.

May, T. and Perry, B. (2016). *Cities and the Knowledge Economy: Promises, Politics and Possibilities*. London: Routledge.

Miao, B. and Lang, G. (2014). A tale of two eco-cities: experimentation under hierarchy in Shanghai and Tianjin. *Urban Policy and Research*, 33: 247–263.

Miao, J., Benneworth, P. and Phelps, N. (eds) (2015). *Making 21st Century Knowledge Complexes: Technopoles of the World Revisited.* London: Routledge.

Morgan, K. (2001). The exaggerated death of geography: localised learning, innovation and uneven development. Paper presented at *The Future of Innovation Studies Conference.* Eindhoven Centre for Innovation Studies, Eindhoven University of Technology. 20–23 September 2001.

Neef, D. (1998). *The Knowledge Economy.* Woburn, MA: Butterworth-Heinemann.

Oakley, A. (2000). *Experiments in Knowing: Gender and Method in the Social Sciences.* Cambridge: Polity.

O'Mara, M. (2005). *Cities of Knowledge. Cold War Science and the Search for the Next Silicon Valley*. Princeton, NJ: Princeton University Press.
Ong, A. (2011). Introduction: worlding cities or the art of being global. In Roy, A. and Ong, A. (eds), *Worlding Cities: Asian Cities and the Art of Being Global*. Oxford: Blackwell, 1–26.
Only Lyon (2014) *Smart City: Experimenting Today for Better City Living Tomorrow*. Lyon: Greater Lyon Economic and International Development Delegation.
Ooi, C.-S. (2008). Re-imagining Singapore as a creative nation: the politics of place-branding. *Place Branding and Public Diplomacy*, 4: 287–302.
Oosterlynck, S. and Gonzalez, S. (2013). Don't waste a crisis: opening up the city yet again for neoliberal experimentation. *International Journal of Urban and Regional Research*, 37: 1075–1082.
Perry, B and May, T. (eds) (2007). Governance, science policy and regions. *Regional Studies*, 41 (Special edition).
Perry, B and May, T. (eds) (2011). Building knowledge cities: the roles of universities. *Built Environment*, 37 (Special edition).
Perry, B. and May, T. (2015). Context matters: the English science cities and visions for knowledge-based urbanism. In Miao, J., Benneworth, P. and Phelps, N. (eds), *Making 21st Century Knowledge Complexes: Technopoles of the World Revisited*. London: Routledge, 105–127.
Pickstone, J.V. (2000). *Ways of Knowing: A New History of Science, Technology and Medicine*. Manchester: Manchester University Press.
Pike, A., Rodriguez-Pose, A. and Tomaney, J. (2006). *Local and Regional Development*. London: Routledge.
Polk, M. (ed.) (2014). *Co-producing Knowledge for Sustainable Cities: Joining Forces for Change*. London: Routledge.
Poovey, M. (1998). *A History of the Modern Fact: Problems of Knowledge in the Sciences of Wealth and Society*. Chicago: University of Chicago Press.
Poster, M. (1990). *The Mode of Information: Poststructuralism and Social Context*. Cambridge: Polity.
Rothenberg, M.A. (2010). *The Excessive Subject: A New Theory of Social Change*. Cambridge: Polity.
Roy, A. and Ong, A. (eds) (2011). *Worlding Cities: Asian Cities and the Art of Being Global*. Oxford: Blackwell.
Sassen, S. (2012). *Cities in a World Economy, 4th Edition*. Thousand Oaks, CA: Sage.
Savitch, H. and Kantor, P. (2004). *Cities in the International Marketplace: The Political Economy of Urban Development in North America and Western Europe, New Edition*. Princeton, NJ: Princeton University Press.
Sayer, A. (2015). *Why We Can't Afford The Rich*. Bristol: Policy Press.
Scott, A.J. (2000). *The Cultural Economy of Cities*. London: Sage.
Scott, A.J. (2014). Beyond the creative city: cognitive–cultural capitalism and the new urbanism. *Regional Studies*, 48: 565–578.
Simmie, J., Sennett, J., Wood, P. and Hart, D. (2002). Innovation in Europe: a tale of networks, knowledge and trade in five cities. *Regional Studies*, 36: 47–64.
Spilhaus, A. (1967). The experimental city. *Daedalus*, 96: 1129–1141.
Stehr, N. (1992). *Practical Knowledge: Applying the Social Sciences*. London: Sage.
Stehr, N. (1994). *Knowledge Societies*. London: Sage.
Vico, G. (1990). *On the Study Methods of Our Time*. Ithaca, NY: Cornell University Press.

Wong, C., Choi, C.-J. and Millar, C. (2006). The case of Singapore as a knowledge-based city. In Carillo, F. (ed.), *Knowledge Cities: Approaches, Experiences and Perspectives.* Oxford: Butterworth-Heinemann, 87–96.

Wyly, E. (2013). The city of cognitive cultural capitalism. *City*, 17: 387–394.

Youtie, J. and Shapira, P. (2008). Building an innovation hub: a case study of the transformation of university roles in regional technological and economic development. *Research Policy,* 37: 1188–1204.

Žižek, S. (2009). *Violence: Six Sideways Reflections*. London: Verso.

4 The urban laboratory and emerging sites of urban experimentation

Simon Marvin and Jonathan Silver

Introduction

In this chapter, we examine the emerging forms of experimentation being undertaken across 'urban laboratories' to consider the significance of these learning spaces in wider forms of technological and sustainability orientated urbanisation processes. A growing body of work seeks to interrogate the relation between these spaces of innovation and new forms of urbanism, infrastructure and technology development in contemporary urban regions (Gieryn 2006; Evans and Karvonen 2011). The basis of these spaces has been recently examined by Karvonen and van Heur (2014: 380) who suggest that 'urban laboratories centre on processes of change and the emergence of new practices and concepts, connecting future visions of cities to the "politics and practices of hope"'. Yet, there has been little systemic examination of the multiple contexts, drivers, financing, partnerships and objectives of the plethora of emerging urban laboratories. In this chapter, we contribute to recent debates by examining the wider socio-technical landscapes and particular configurations of these experimental spaces to better understand the multiple ways in which they are reconfiguring diverse urban-institutional contexts. The chapter draws on a survey and database of over 70 of these urban laboratories in the UK and internationally.

In the next section, we examine the role of urban laboratories in fostering experimentation, innovation and learning about new forms of urbanism, technological change and wider urban development. We then outline the results of mapping these urban laboratories across the UK and internationally in the third section via a survey-generated database. In the fourth section, we provide an overview of the different styles of urban experimentation being undertaken in terms of the multiple logics, settings, foci, activities and temporalities. We then interrogate the typology and reflect on the practices and priorities of urban laboratories. We conclude by reflecting on how these urban laboratories are configuring particular landscapes of experimentation, the future trajectories that are being shaped and how they relate to wider urban (technological) transformation.

The emergence of the urban laboratory

Cities are increasingly being governed through forms of experimentation that seek to generate new urban conditions, often orientated around future sustainability concerns through technological development, new forms of organisation and low-impact living (Bulkeley *et al.* 2014). An urban laboratory is a space designed for interactions between an urban context and a research process to create the conditions for this experimentation, creating new forms of urbanisation through testing, developing or applying social practices or a technology to a building or wider infrastructure system. Karvonen and van Heur (2014: 389) argue that 'these spaces of innovation and change provide a designated space for experimentation where new ideas can be designed, implemented, measured and, if successful, scaled up and transferred to other locales'. They are also organised in quite different ways – whether geographically, through the ways of working, purpose and orientation and of course how they are governed and managed and often dependent on the factors shaping their emergence.

A range of socio-technical landscapes (Hodson and Marvin 2010) linked to a series of shifting urban problematics are prompting the emergence and development of these urban laboratories across the university sector and beyond. These include the growing importance of urban agendas around low carbon (Bulkeley 2005; Hodson and Marvin 2011), smart urbanism (Greenfield 2013; Luque-Ayala and Marvin 2015) and the wider positioning of urban regions in the global economy (Sassen 2012). Such dynamics are unfolding both in the UK and internationally to become key governance and knowledge production spaces in addressing uncertain and indeterminate urban futures (Zeidermann *et al.* 2015). Universities, companies and other urban actors face a range of external and internal pressures to engage with these future-driven knowledge economy agendas (Perry and May 2011) and the urban laboratory provides a response to such twenty-first century urban imperatives. Linked to these growing pressures are a range of opportunities that are emerging as cities become test beds for experimentation through shifting forms of urban governance (Bulkeley *et al.* 2014). These include new funding and financing directed to the knowledge economy across the UK and further afield from a series of different sources and social interests both within the university sector, national government and municipalities but also increasingly across the commercial sector. The need for these various governance actors to link to knowledge production and practices is vital to both public visibilities of addressing key global socio-environmental concerns and the strategic direction of future preparedness. A central part of this emergence is predicated on the landscape of technological innovation to address the twenty-first century challenges of urbanisation and the urban laboratory provides a space in which transformations in innovation and technology pathways can be realigned to address this agenda (Evans and Karvonen 2014). Furthermore, this somewhat elusive laboratory landscape is being shaped by developing partnerships between universities and large businesses that are seeking to bring together knowledges concerning financing, expertise and increasingly, techno-market opportunities (Perry and May 2011).

The importance of the 'impact agenda' has meant that universities are exploring multiple ways to demonstrate how research is able to generate benefits for wider society and show the relevance of the university to wider urban life (May and Perry 2006). Linked to this impact agenda is the growing awareness of the need to develop a broad sustainability brand for the university, with a range of initiatives that are directly related to the running of the university operations and the role of the campus as a test bed for different experiments. Whilst these dynamics are shaped by key university actors, they can provide space for a series of student and community partnerships that move beyond the campus as an experimental site. Such dynamics are often linked to the role of the spaces in seeking to create a public-facing visibility that the university can mobilise to appeal to an external audience of potential students, businesses and grant giving organisations (Perry and May 2011).

Alongside the emergence of such spaces across universities, urban laboratories have become mobilised across the commercial sector. This is particularly evident in energy and ICT (information and communications technology) companies seeking to respond to new digital, smart and low carbon market opportunities. Such commercial spaces provide the conditions to test, experiment and develop new innovations for the market with particular urban spaces being used as (promotional) test beds to export and upscale into other urban contexts. These labs often also provide a showroom for new products or for the company's sustainability credentials and are opened permanently or during particular staged moments to wider publics. The Crystal in London is perhaps the best example of this showroom/public interaction space. It was specifically designed as an outward facing space that is focused more on promoting the Siemens brand than testing new technologies (which takes place at other sites).

Whilst the idea of urban laboratories remains closely associated with the university sector, the term is travelling further afield to include communities and various other scales of the urban region. As Evans and Karvonen (2014: 427) suggest, 'cities are racing to attract scientists and companies with scientific infrastructure to enhance their economies and improve their international reputations'. These new forms of public–private partnerships are travelling in a number of different directions predicated on the key social interests involved in the space and providing a shifting socio-technical landscape of potential possibility and promise.

Mapping urban laboratories

We undertook a mapping exercise involving both desktop research (using secondary materials to identify and collect information on the urban laboratories) and follow-up interviews with 15 representatives. The aim of the exercise was to construct a database to summarise the growing number of such initiatives. Over 70 urban laboratories were identified both at the UK level and at a wider international scale (see Figures 4.1 and 4.2), with a focus on those broadly working across sustainability. This was undertaken by engaging with various networks engaged in these issues, with the use of keyword searches (lab, living lab, low carbon lab, technology lab and so forth) and through existing knowledges of and collaboration with such spaces. We sought to include as many of these initiatives as possible to develop a comprehensive database that could provide a snapshot of a shifting, and at times accelerating, landscape of urban experimentation.

Key

A1 Queens Building/Living Lab
A2 Salford Energy House
A3 European Bioenergy Research Institute
A4 Energy Business Research Lab
A5 Future Energy Labs
A6 Oxford Road Corridor
A7 Intel Collaborative Research Institute for Sustainable Connected Cities
A8 Sustainable Innovation Lab (SILAB)
A9 CREST
A10 Smart Grid Lab
A11 SmartLIFE Low Carbon Centre
A12 The Centre for Efficient and Renewable Energy in Buildings
A13 The Energy Systems Integration Lab
A14 Carbon Co-op
A15 Brighton Waste House
A16 Inheritable Futures Laboratory [IF:LAB]
A17 HP Labs
A18 Living Lab for Sustainability
A19 Biospheric Foundation
A20 Real Food Wythenshawe
A21 The Birmingham Urban Climate Lab (BUCL)
A22 Manchester Environmental Resource Centre Initiative (MERCI)
A23 Northumbria Design Social Innovation and Sustainability Lab
A24 Imagination Lancaster
A25 Camp for Climate Change
A26 Suttle Lab
A27 Sustainability Hub
A28 Keele Hub for Sustainability
A29 Pedal Powered Cinema
A30 Edinburgh Centre for Carbon Innovation
A31 Living Lab Coventry
A32 The Campus as Living Lab
A33 Living Lab
A34 The Crystal
A35 Incredible Edible
A36 LILAC
A37 Zero Carbon Hub
A38 Centre for Alternative Technology

Figure 4.1 Map of UK urban laboratories.

Key

G1 Flemish Living Lab Platform (FLLP)
G2 Malmo New Media Lab
G3 Urban LIving Lab (ULL)
G4 EURANLAB
G5 Arcosanti
G6 North Desert Village
G7 Elmer Avenue
G8 Tianjin Eco-city Low Carbon Living Lab
G9 Victorian Eco Innovation Lab
G10 City Lab: Climate Change
G11 Carbon Neutral Energy Solutions Labratory
G12 Carbon Lab
G13 Sustainable Neighbourhood Lab
G14 Experiment Days
G15 Christiania
G16 Smarter Infrastructure Lab
G17 Infrastructure Network Lab
G18 Green Futures Lab
G19 Resillient Infrastructure Network Lab
G20 Urban Design Lab
G21 Community Infrastructuring

Figure 4.2 Map of global urban laboratories.

G22 The SCI Energy Lab
G23 Stellenbosch Sustainability Institute
G24 The Urban Action Lab
G25 Studio for the Study of Social Materialities in the Global South
G26 Hyderabad Urban Lab
G27 Digital Home Energy Management Systems (DEHEMS)
G28 Smart Grid Test Project Brasov City
G29 Pecan Street Project
G30 Masdar City
G31 Centre for Innovation, Testing and Evaluation

The interviews focused on understanding the key motivations, drivers and activities of each urban laboratory and were conducted over the phone and through visits to these spaces. Whilst many of these can be recognised explicitly as 'laboratories', some remain less defined in relation to these terms but have been included to consider how we understand such spaces. Data collection took place in June and July 2013, with a revision later in the year. Whilst the use of the internet has some limitations, this was supplemented with interviews and literature reviews to give a broader understanding of the findings.

This survey work was followed by the development of a database predicated on a typology of urban laboratories to interrogate the 70+ examples collected. This has meant examining a series of factors that shape the logics involved in the conception of how these spaces are designed, the types of activity being undertaken, the focus of these activities, and the temporalities of these spaces. The typology is meant to act as a way to examine the different ways in which urban laboratories emerge and the competing drivers and dynamics that shape such initiatives. Whilst the work has sought to characterise each of the urban laboratories through these typologies, it should be noted that many of these spaces are co-produced through the relationalities of overlapping, competing and complimentary agendas, activities, settings and foci, illustrating the often 'messy' socio-technical landscape in which they emerge.

A typology of forms of experimentation

In the next section, we report on the analysis of the 70+ urban laboratories in the survey. We suggest that a typology can be built around the identification of a number of different forms of experimentation.

Lab logics

Logics refers to the drivers that shape the purpose and outlook of urban laboratories and frames the rationale for the emergence of the space, the expectations of its outputs and the role that it may play in urban experimentation and transformation. The first logic is that of economic growth in which the formation of the urban laboratory is predicated on finding ways to create new economic opportunities, explicitly centred on the commercialisation of sustainability research and opportunities for future capital accumulation. This neoliberal model of experimentation around sustainability concerns is perhaps best illustrated in the Masdar City development in the United Arab Emirates as an 'eco-city' built through ongoing testing and experimentation of new forms of green urbanism and a utopian rhetoric of urban futures. These logics are not only found in private companies but increasingly in public–private partnerships between universities and the commercial sector. Whilst initial urban laboratories were focused on technology development, this is broadening to include a range of different activities increasingly linked to urban growth strategies. This suggests that these experimental spaces are becoming central to emerging twenty-first century urban development strategies.

The second logic is that of education/knowledge production in which the establishment of urban laboratories is predicated on knowledge production concerning urban futures and orientated around educational purposes. For example, the European Bioenergy Research Institute in the West Midlands (UK) focuses on academic research to examine, test and generate knowledge around this important future energy source. Whilst the knowledge may be used for commercial application, the logic of the space is not restricted to this motivation. A number of these spaces also offer training and other learning opportunities for wider publics, providing education to produce a highly skilled workforce, support communities and offer opportunities for students and staff to engage with this agenda.

The third logic is that of techno-orientation in which urban laboratories are set up to generate new technologies through intensive modelling, testing and application of softwares, sensors, apps and other smart city infrastructures. For example, the Salford Energy House provides a facility for testing new retrofit technologies and incremental upgrading. Such spaces have attracted significant financial investment from a range of partners. This funding emerges from governments concerned with binding climate change agreements, research councils and commercial building organisations.

Post-capitalist living forms the fourth identified logic, generating the creation of urban laboratories that seek to envisage a steady-state transition beyond the current modes of a growth economy that has damaged both people and planet. For example, the Manchester Environmental Resource Centre initiative (MERCi) in Manchester is both a community facility and an active test bed for non-market responses to socio-environmental imperatives. These are often built through low-cost efforts and peripheral in relation to the higher profile techno-orientated labs, full of rhetoric and smart city visions of potential futures. Whilst this post-capitalist logic may interact with other logics, the urban laboratories established through this also provide a challenge to particular visions of the urban future through a DIY, community-driven urbanism.

Lab settings

Setting refers to the institutional context of urban laboratories and is a key factor to how they emerge, stabilise and grow. Whilst urban laboratories have traditionally been found on university campuses, they are now common in a range of different settings, particularly partnerships that span beyond the immediate university environment. These settings tend to reflect the key social interests involved in the shaping and management of these experimental spaces. The first setting is that of the university campus, accounting for over 50 per cent of the total urban laboratories in the database. They are funded through the academic sector but incorporate a range of partnerships with various different actors involved around sustainability issues (Trencher *et al.* 2014), such as Imperial College's Energy Research Business Lab. Universities provide the central coordinating partner in the emergence of the space and are actively involved in the management and direction of research and development activities. This setting is the most common for urban laboratories and remains vital to the growing visibility of such initiatives across urban landscapes.

The second setting is commercial, accounting for 10 per cent of the total, where urban laboratories are funded and staffed via private corporations with the aim of generating future markets. An example of this is the HP Labs in Bristol. These urban laboratories are often linked to university activities in various forms but are situated within a commercial setting and are dominated by both large energy and ICT companies that seek to test, explore and sell these new technologies, infrastructures and accompanying operations. The emphasis in these urban laboratories is often on practical outputs that have some commercial utility.

The third setting is built around the notion of partnerships between multiple actors and accounts for 30 per cent of these spaces, often including universities and which may have a range of logics that shape the direction of the urban laboratories. Such partnerships are increasingly common around technology development in which large amounts of financing are needed to sustain the activities of the space and the multiple logics which are embedded in these activities. As universities become increasingly strategically aligned to the needs of industry partners, this setting is becoming a key part of the landscape of these spaces, such as the Oxford Road Corridor in Manchester (see Evans and Karvonen 2014).

A fourth setting, accounting for just 5 per cent of urban laboratories, is based on the community or alternative space and is based on post-capitalist logics in which the setting is predicated on ways to radically rethink the trajectories of relations between capitalism, people and planet. Whilst such urban laboratories might be set up by university staff and student activities, the setting might equally be produced through community activist activities that seek to explore alternatives modes of neighbourhood development and regeneration. One such example of this setting is the LILAC development in Leeds (see Chatterton 2013). Unlike the commercial setting, the placing of urban laboratories in these contexts produces a range of outputs that may contribute to alternative imaginings of urban futures.

Finally, the fifth setting is based on hybridity and built around a mix of different types of settings and partners; this setting makes up 5 per cent of total laboratories. Whilst all labs may be classified as such, there are a number of examples of lab spaces that explicitly function as a hybrid of multiple settings and act as a space in which different actors can interact and collaborate. For instance, the Pecan Street Project in Austin, Texas draws together partners including the City of Austin, Austin Energy, The University of Texas at Austin, the Austin Technology Incubator, the Greater Austin Chamber of Commerce and the Environmental Defence Fund in a collaboration to reinvent community energy networks.

Lab activities

Activity refers to the type of work that urban laboratories are undertaking, as they are all set up to explore ways to generate different types of research and development, including data, knowledge, technology application, experimentation and so on. The first set of activities is based around research, a core feature of many urban laboratories and in around 20 per cent form the only activity. Such activities often act as a way to bring together a research cluster within a university setting, perhaps even without a physical space from which to develop new approaches to sustainability. They include, for example, the Future Energy Labs being run by Imperial College London and the Hyderabad Urban Lab in India. These knowledge spaces are emerging as a key route for university departments and cross-departmental initiatives to bring together various research strands and position and brand their research to wider publics.

The second set of activities is established around research and development. Whilst a number of urban laboratories focus on research-only activities, 40 per cent are also involved with undertaking development of new technologies in the built environment, design and architectures, through activities to test, upscale and explore these new applications. Here there is a clear link between knowledge production and wider urban and more traditional understandings of laboratory activities, an example being the Carbon Lab at the University of Queensland.

The third set of activities is focused on testing and makes up around 25 per cent of urban laboratories. These urban laboratories are focused on generating and piloting new ideas for application in the urban landscape and this can include focusing on smart cities, low carbon living or new technologies. In these urban laboratories, there seems to be less emphasis and focus on research; the learning taking place is something that happens incrementally through experimentation and refining ideas or through the running of simulations that provide data that is often sent or sold to external partners for analysis or to improve technology development. Much of the research and analysis emerging from these spaces thus takes place off site.

Finally, the fourth set of activities is built around training and education. These urban laboratories are focused on or incorporate activities to develop professional or public knowledge around a series of urban issues (sustainability, smart/low carbon) and include around 15 per cent of the total database. Activities include more formal education such as undergraduate teaching, professional training, consultancy and community training. Activities are focused on providing behavioural change type training, the development of specialist skills concerning technologies and the wider low carbon economy, or the context in which such changes are taking place.

Lab foci

Focus describes the strategic direction of urban laboratories and the fields of intervention they are operating across. First, new technologies in which the focus in the urban laboratories is predicated on developing new technologies at a range of scale from the household to the city accounts for 35 per cent of total labs. These technologies may be linked to wider research that is not necessarily framed via an urban lens with the urban laboratories addressing and translating these technologies within an urban context. Such activities form a key characteristic of many of the urban laboratories in the mapping exercise and include the Smart Infrastructures Lab in Pennsylvania.

Second, the labs can be focused on the urban landscape through developing a range of interventions across the built environment of cities, such as the Sustainable Neighbourhood Lab in Boston, MA. Whilst many of these urban laboratories are centred on campuses, there has been an increasing blurring of the lines between the campus and the wider city. These urban laboratories are thus externally focused, not just out to the wider city but to wider discourses that the city itself projects to the world (smart, low carbon, innovation, global) as part of its place branding and work on inward investment. Whilst they only account for 5 per cent of current urban laboratories, we expect these to grow over the next few years.

The third area of focus is centred on building retrofit and exploring ways to retrofit buildings using new technologies, specifically aimed at energy efficiency measures, low carbon technologies and a range of complimentary smart technologies; this area makes up around 15 per cent of urban laboratories, including the Zero Carbon Hub in Milton Keynes. There are a number of examples of the establishment of particular housing types and the testing and experimentation that can be developed in these spaces, together with work around commercial building retrofit. Alongside the testing of new ways to retrofit in lab type conditions, there are a number of spaces that are also themselves being retrofitted and act as live test beds for these experiments.

The fourth area of activities comprising around 8 per cent of the spaces is based on the low carbon economy and is predicated on developing economic outputs, measurements or skills needed for urban low carbon transitions, such as the SmartLife Low Carbon Centre in Cambridge. As Hodson and Marvin, (2007: 303) suggest, 'The climate-change agenda is reinvigorating a need to "cultivate new techniques of governance" for urban sustainability', and low carbon is central to discourses about future cities. The need to develop the range of economic activities beyond a focus on technology development is playing a role in the formation and pathways of urban labs. Such activities include business consultancy, energy efficiency auditing, retrofit and training/skills development.

The fifth area of activities is focused on food production, with the development of local food solutions through new low carbon technologies, systems and thinking, and accounts for 7 per cent of urban laboratories. Initiatives such as the Biospheric Foundation in Salford reflect the increasing agenda around growing food locally and sustainably, and the visibilities around vertical farm systems and new forms of technology-based growing.

The sixth area of activities is based on sustainability and may have been established through a broader agenda emerging from a range of concerns; it accounts for around 20 per cent of urban laboratories. These activities are often associated with social responsibility agendas or are played out in environments that have been established for some time and are witnessing a shift or transition into a range of different logics and activities, such as the Sustainability Hub in Hull.

Finally, there are activities that directly address climate change agendas that may incorporate low carbon dynamics but are also working on these issues within a wider context. This focus, accounting for 10 per cent of urban laboratories, is developed either through supporting new technologies and research methodologies in capturing data or supporting policy development by providing spaces to draw in climate science, modelling and simulations, as seen at the Suttle Lab at Imperial College London.

Lab temporalities

Temporality refers to the timescales of urban laboratories and is an important consideration in the development of these spaces, showing a range of stages from temporary and uncertain, in which the space is still being defined, to more long-term temporalities that often have a more stable focus and range of activities.

First, long-term urban laboratories are funded and well established and their role in the wider city is well embedded. Likewise, their functions and activities are well defined. This temporality accounts for over 70 per cent of total urban laboratories.

Second, the urban laboratory as a temporary and short-term intervention accounts for 20 per cent of the total database and is characterised by experiments with a particular focus. These temporary spaces may be associated with a range of different settings that seek to provide a period of time for participants to gather and share knowledge, collaborate and build links and networks.

Third, is the category of uncertain, which accounts for 10 per cent of urban laboratories. The future of their activities is provisional. They may be seeking funding or may be set up for the long term but are dependent on future financing. As such, the focus, activities and agenda may be going through different iterations and being contested by different actors involved in the space.

Trends and patterns in the growth of urban laboratories

If, as Evans and Karvonen, (2014: 427) posit, 'The appeal of the urban laboratory as a mode of governance lies in its potential to transform the economic and social landscape,' then it is important to consider the type of transformations taking place. Based on ongoing engagement with a number of these spaces, together with the research conducted for this chapter and extensive communication with a range of networks working on similar issues, we suggest a number of emergent trends and patterns.

The range of urban laboratories identified in this review suggest a number of important dynamics and considerations about these increasingly visible spaces of experimentation and future-driven testing. The first trend and pattern worth considering is the trajectories of urban laboratories and the role they play in validating and legitimating particular forms of (often neoliberal) innovation and transition pathways. It is worth noting that whilst the logic of economic development and growth is important in many of these spaces and is helping to facilitate industry development, there are also a series of experimental spaces that seek to explore alternative models around urban futures that suggest radical transformations of the socio-technical configurations of cities. The relationships between and across these spaces will be important in the future role of urban laboratories and the connections they have with wider society.

Second, the trends and patterns of the selected urban laboratories suggests important considerations about the future of university research. This is increasingly public facing, building links to industry and seeking to develop both visible research impact together with positioning themselves within increasingly competitive environments for student and staff recruitment. A key question emerging from this process is to what extent urban laboratories produce or reinforce such dynamics as well as the marketisation of university research and neoliberalisation of the academy (Levidow 2002). A third trend taken from the database suggests that emerging low carbon/smart technologies are being developed in the UK and show the important role urban laboratories are playing in experimenting and testing in real-world situations.

Fourth, it is possible to suggest that urban laboratories are playing an increasingly central role in shaping particular urban transition pathways. This is particularly visible in cities with clear policy development concerned with low carbon agendas.

Fifth, the uneven geographies of laboratorisation and the flows of knowledge, benefits and technologies will become increasingly important in relation to the types of interest that are producing knowledges about urban futures and transitions. The urban intermediaries involved in urban laboratories are varied. They range from large corporations through to post-capitalist activists and show how these actors are shaping particular cultures, rationalities and logics concerning these spaces.

Sixth and linked to the role of these urban laboratories in shaping urban transition pathways is the role these spaces play (or not) in local politics and the type of urbanisms being imagined, speculated and upscaled as new forms of potentially uneven urban futures. Many of the discourses being produced are predicated on notions, aspirations and claims of technologically advanced 'SMART' cities (Greenfield 2013; Kitchen 2014) that will control flows of resources, harness renewable energy sources and make low carbon living an increasingly normalized practice for the population. Yet many of these narratives fail to link the low carbon agenda to other long-standing urban issues such as poverty, deprivation and splintered urbanisms (Graham and Marvin 2001). As such, the political role of these urban laboratories becomes an important factor in how they develop and the role they play in wider debates and contestations over urban futures.

Conclusions

In this chapter, we have sought to provide a review of the emerging forms of experimentation taking place across urban laboratories. We have shown through the typology the many different motivations, organisations, purposes and potential implications of these spaces. The study suggests some important reflections on the wider political-economic roles and the socio-political organisation of these experimental spaces in wider urbanisation processes. Reflecting on the findings, we find there are two sets of issues for further research.

The first concerns how we might start to understand the particular political and politicised roles that urban laboratories perform in wider urban change. There are three competing roles here that urban laboratories might powerfully relate to. The first is their role in supporting the city as an 'economic engine' in which the space is linked to the wider economic growth objectives of the neoliberal urban project, increasing ground rents and a range of city metrics (Theodore *et al.* 2011). The urban laboratories in this context are particularly focused on a narrow set of growth objectives – enhancing the uptake of products and services in infrastructure networks, accelerating the efficiency of infrastructure services, supporting the ecological modernisation of infrastructures through incremental and technical change, and developing pockets of innovation. Here, the role is primarily one of economic boosterism rather than supporting more radical change and transition in socio-technical infrastructures, a shift we would suggest in technological rather than societal relations. The second role would see the urban laboratories as part of wider processes of 'global urban competition' (Jessop 2002). Here, the urban laboratories would be understood as a critical urban economic asset that has the capacity to compete for scarce funding resources, provide evidence of an urban exemplar or flagship project, and act as a resource in building networks with the university sector and corporate partners. In this context, we might see these urban laboratories as currently very fashionable yet largely irrelevant exemplars of competitive urbanism. Such experimental spaces are failing to fundamentally address the issues of renewing urban infrastructures yet support 'world city' aspirations and claims of civic innovation in new post-networked forms of splintered urbanism (Courtard and Rutherford 2011). The final role is urban laboratories producing an 'enhanced urban capacity'. In this role, the successful laboratory may be able to open up new research and policy agendas and enrol new and previously marginalised or excluded social interests in the innovation process. Through these processes, critical reflection and reflexive learning may emerge to develop the potential for progressive urban change through new technological forms. Further research will be necessary to establish what priorities are driving the development of urban laboratories in different urban contexts and what their potential is to produce socio-material change in urban infrastructures that challenges current splintered urbanisms (Graham and Marvin 2001) rather than reflecting current urban power relations.

The second set of issues concerns the social organisation of urban laboratories and in particular the ways in which different sets of priorities and social interests can be brought together in specific lab configurations. While there is evidence of a growth in urban laboratories – or at least a renaming and repositioning of existing initiatives through this term – there is also a more diverse set of configurations. There are three types in practice that would be worth subjecting to critical comparative research. The first are Urban Living Labs (ULLs) that are new collaborations devised to design, test and learn from social and technical innovation in real time – a key example is the Malmö Living Lab, EURBANLAB. A European network provides support for the development of an approach that is primarily designed to develop a 'new innovation strategy that increases the chance your products become a commercial success' (ENOLL 2015). The ULL enrols end users into the innovation process and develops international networks for the transfer of technologies. Second, 'Urban Transition Labs' (UTLs) are settings in which collaborators deploy and carefully observe sustainability transitions, as for instance in the example of the MUSIC network of five cities (see Nevens *et al.* 2013). UTLs follow a carefully designed process of experimentation and learning around the limits and potentials of urban transitions. Finally, Urban Knowledge Arenas (UKAs) are collaborative platforms to provide knowledge exchange for shared urban priorities, with the Mistra Urban Futures (2015) an example of this type of laboratory. UKAs are more concerned with developing new approaches to the development of knowledge on a wide range of urban issues – not solely focused on innovation and infrastructures (COST C20 2010). Consequently, there are a much wider range of laboratories in an urban context and there is thus a need for greater sensitivity in how the different configurations relate to different types of urban experimentation – some more product oriented, others more systemic oriented towards transitions and other concerned with the new forms of urban knowledge exchange.

References

Bulkeley, H. (2005). Reconfiguring environmental governance: towards a politics of scales and networks. *Political Geography*, 24: 875–902.

Bulkeley, H., Castán Broto, V. and Edwards, G. (2014). *Climate Experiment Cities: An Urban Politics of Sociotechnical Transition*. London: Zed Books.

Chatterton, P. (2013). Towards an agenda for post-carbon cities: lessons from LILAC, the UK's first ecological, affordable cohousing community. *International Journal of Urban and Regional Research*, 37: 1654–1674.

COST C20 (2010). *Urban Knowledge Arenas*, Final report, COST (European Cooperation in Science and Technology).

Coutard, O. and Rutherford, J. (2011). Post-networked cities: recombining infrastructural, ecological and urban transitions. In Bulkeley, H., Castán Broto, V., Hodson, M. and Marvin, S. (eds), *Cities and Low Carbon Transitions*. London: Routledge, 107–125.

ENOLL (2015). *European Network of Living Labs Knowledge Centre*. [Online]. Available: http://knowledgecenter.openlivinglabs.eu [15 August 2015].

Evans, J. and Karvonen, A. (2011). Living laboratories for sustainability: exploring the politics and epistemology of urban transition. In Bulkeley, H., Castán Broto, V., Hodson, M. and Marvin, S. (eds), *Cities and Low Carbon Transitions*. London: Routledge, 126–142.

Evans, J. and Karvonen, A. (2014). 'Give me a laboratory and I will lower your carbon footprint!' – Urban laboratories and the governance of low-carbon futures. *International Journal of Urban and Regional Research*, 38: 413–430.

Gieryn, T.F. (2006). City as truth-spot: laboratories and field-sites in urban studies. *Social Studies of Science* 36: 5–38.

Graham, S. and Marvin, S. (2001). *Splintering Urbanism: Networked Infrastructures, Technological Mobilities and the Urban Condition*. London: Routledge.

Greenfield, A. (2013). *Against the Smart City*. New York: Do projects.

Hodson, M. and Marvin, S. (2007). Understanding the role of the national exemplar in constructing 'strategic glurbanization'. *International Journal of Urban and Regional Research*, 31: 303–325.

Hodson, M. and Marvin, S. (2010). Can cities shape socio-technical transitions and how would we know if they were? *Research Policy*, 39: 477–485.

Hodson, M. and Marvin, S. (2011). Making low-carbon England and Wales. *Town and Country Planning*, 79: 388–392.
Jessop, B. (2002). Liberalism, neoliberalism, and urban governance: a state–theoretical perspective. *Antipode*, 34: 452–472.
Karvonen, A. and van Heur, B. (2014). Urban laboratories: experiments in reworking cities. *International Journal of Urban and Regional Research*, 38: 379–392.
Kitchin, R. (2014). The real-time city? Big data and smart urbanism. *GeoJournal*, 79: 1–14.
Levidow, L. (2002). Marketizing higher education: neoliberal strategies and counter-strategies. In Robins, K. and Webster, F. (eds), *The Virtual University? Knowledge, Markets and Management*. Oxford: Oxford University Press, 227–248.
Luque-Ayala A. and Marvin S. (2015). Developing a critical understanding of smart urbanism? *Urban Studies*, 52: 1–12.
May, T. and Perry, B. (2006). Cities, knowledge and universities: transformations in the image of the intangible. *Social Epistemology*, 20: 259–282.
Mistra Urban Futures. (2015). *Mistra Urban Futures*. [Online]. Available: http://www.mistraurbanfutures.org [15 August 2015].
Nevens F., Frantzeskaki, N., Gorissen, L. and Loorbach, D. (2013). Urban transition labs: co-creating transformative action for sustainable cities. *Journal of Cleaner Production*, 50: 111–122.
Perry, B and May, T. (eds) (2011). Building knowledge cities: the roles of universities. *Built Environment*, 37 (Special edition).
Sassen, S. (2012). *Cities in a World Economy, 4th Edition*. Thousand Oaks, CA: Sage.
Theodore, N., Peck, J. and Brenner, N. (2011). Neoliberal urbanism: cities and the rule of markets. In Bridge, G. and Watson, S. (eds), *The New Blackwell Companion to the City*. Oxford: Wiley-Blackwell: 15–25.
Trencher, G., Yarime, M., McCormick, K.B., Doll, C.N. and Kraines, S. B. (2014). Beyond the third mission: exploring the emerging university function of co-creation for sustainability. *Science and Public Policy*, 41: 151–179.
Zeiderman, A., Kaker, S. A., Silver, J. D. and Wood, A. (2015). Uncertainty and urban life. *Public Culture*, 27: 1–19.

5 Virtual city experimentation

A critical role for design visioning

Chris Ryan, Idil Gaziulusoy, Kes McCormick and Michael Trudgeon

Introduction: virtual experimentation and the transformation of cities

Sengers *et al.* (Chapter 2, this volume) define an experiment as 'an inclusive, practice-based and challenge-led initiative designed to promote system innovation through social learning under conditions of uncertainty and ambiguity'. Here, we introduce the idea of 'virtual city experimentation': a design approach to catalyse action in the context of rapidly emerging disruptive challenges to the fabric and life of cities. In the meaning we give to the term, the concept of 'virtual' experimentation owes little to the contemporary association with 'digital', or 'on-line'. Independent of any technology used in the process, 'virtual' here speaks about evoking the critical human ability to conceptualise alternative realities, to imagine and to explore *in the mind* other sets of relationships (social, physical, technological) than those currently evident in the lived-in world.

Any realistic assessment of the multiple challenges of climate change suggests that human society is facing a period of revolutionary transformation, as significant as any in history. Compared to past revolutionary shifts, such as the first Industrial Revolution that lasted 80 to 100 years (Landes 1972), the response to anthropogenic climate change has to be rapid as well as fundamental, involving the transformation of technologies as well as lifestyles and social practices. If this revolutionary period happens without major social dislocation and contestation, this will be the result of creative innovation, an acceptance of uncertainty and ambiguity as a condition of knowledge, and an engagement with the idea and value of experimentation as a social, cultural and economic objective. Here we explore the possibilities of 'virtual' experimentation as one answer to the looming societal question: how can we stimulate enough experimentation, rapidly enough, to deal with the pace as well as the scale of change?

The ability to envisage something new and to 'test' its possibilities before bringing it into being, might well define what it is to be human;[1] it certainly defines the process of design and the skills of designers. The projects explored in this chapter emerged from design-based research and education programmes; virtual city experimentation builds on one essential feature of all design activity – the visioning of new potentialities and concepts. Two programmes of design activity at the University of Melbourne within its Victorian Eco-innovation Lab (VEIL) have explored and refined a methodology for virtual experimentation involving design visioning aimed at stimulating rapid innovation. There is a 6-year programme known as 'Eco-Acupuncture' (EcoA) and a 4-year national city programme known as 'Visions and Pathways 2040' (VP2040). Both programmes bring together environmental researchers, academics, masters students and professional designers with the aim of accelerating innovation in response to the challenges of climate change. The focus is on urban life and the city, reflecting a growing consensus that this is probably the critical context in which the social, cultural and technological transformation of a 'post-carbon' economy will be negotiated.

Climate, cities and the need for rapid transformation

We are almost halfway through the period referred to as 'the critical decade' (Hughes and Steffen 2013), the time in which our decisions and actions on climate change will determine the success or failure of a post-carbonaceous transition to avoid severe implications for global society. This is not just a technological issue (though that is the way it is often cast) any more than it is solely a social, economic or governance challenge; addressing climate change requires fundamentally different systems, structures and practices in all those areas combined. The response to climate change will also have to embrace both mitigation and adaptation as a simultaneous objective; the impact of the atmospheric carbon and heat already in the global system means that the climate is changing and will continue to do so for a long time regardless of the pace of decarbonisation.

At the strategic level, it is increasingly recognised that the focus for action in this decade will be *cities*. More than half the world's population now reside in cities and the contribution to global greenhouse gas production from cities is estimated at 75 per cent, even though they occupy only around 2 per cent of the global land area (Satterthwaite and Dodman 2009; UNEP 2011; Hajer and Dassen 2015). Nearly half of the world's cities are considered to suffer from the direct effects of changing climate (UNEP 2011). The critical challenge for cities in this period of climate transformation is that they are complex adaptive systems with significant embedded dependencies built-in over the years of their construction.

It is useful to think of the post-carbonaceous transformation of the city as a process of 'dis-embedding' all of its energy underpinnings: decoupling resource and energy consumption from economic development and, at the same time, switching to non-carbonaceous energy sources. That is challenging enough, but the transformation challenge for cities does not stop at their energy system; the infrastructure and morphology of any city has been shaped (directly and indirectly) by its prevailing climatic conditions. There is mounting evidence that changes in climate patterns and extreme weather events can introduce new vulnerabilities for existing city infrastructure (UNEP 2011; Stone 2012). As climate patterns move beyond their historical envelope and as energy production and consumption is transformed, the *infrastructures of provision* for cities – the systems on which the life of the city depends (e.g. energy, water, food, transport, shelter, waste, products and services and information) – will have to change.

Historical relationships between the city and nature are also under pressure; cities have impacts well beyond their borders, shaping local, regional and continental production systems and transforming landscapes, natural ecosystems and even weather (Marcus and Colding 2011; Stone 2012). Some of those effects stem from what is termed the 'heat island effect' created by the absorption and re-radiation of heat of the sun from the dark surfaces of the city. This city-heat interacts with rising summer temperatures and heatwaves, placing additional pressure on infrastructure (Stone 2012). There is increasing rejection of the conception of the city as a 'refuge from nature' (Kareiva and Marvier 2007; Grimm *et al.* 2008), acknowledging both the bio-physical dependence of the city on natural ecosystem services and psycho-social dimensions of human interaction with nature. Approaches to the transformation of cities include some that start from the core idea of dissolving the boundary between city and country (Beatley 2011; Beatley and Newman 2013).[2]

Importantly, the interest in cities and their place in action on climate change extends beyond their contribution to the global problem and their vulnerability. When it comes to *agency* – the interest, willingness and ability to make changes – there are critical characteristics of cities that appear to offer hope for transformation. Cities (and networks of cities) are active in adopting reduction targets and investing in programmes to reach them (McCormick *et al.* 2013; Kautto and Ryan 2015). The contribution of cities to national economies in most nations is very significant; they can be instrumental in generating the political will and the innovation and creativity necessary for the transition to a post-carbon economy (Bettencourt *et al.* 2007; Bettencourt and West 2011; Glaeser 2011; Katz and Bradley 2013). The (bio)physical challenges to cities have to be viewed in the light of the fundamental *cultural* role of the city; cities are a cultural invention, they result from human social and cultural needs – for association, belonging and exchange, but also for the display and structuring of relationships of power. As cities develop, these social and cultural relationships become intermingled with the layering of the physical form of the city, reproducing systems of meaning that underpin institutional and individual practices and ideas of nature and the social order (Ryan 1985; Hajer 1995; Harvey 2012; Ryan 2013b; Hajer and Dassen 2015). Given that the transformation required for a sustainable future will involve significant changes in patterns of consumption for urban citizens, understanding the ways in which existing patterns of consumption are embedded and reproduced in the urban context will be critical, particularly the interplay between structurally determined consumption (such as car dependency or heating and cooling of poorly insulated dwellings) and socially, culturally or emotionally based consumption learnt as part of life in an urban community.

The dimensions of transformation and the value of experimentation

The tight and embedded interconnections between the critical systems of provision of the city and patterns of living (including consumption) means that trying to re-engineer the city one sub-system at a time is bound to fail because new, often unpredicted, problems are likely to arise in another sub-system. Ultimately, the transformation of cities requires a (rapid) transition from one *set* of socio-cultural-technological-physical systems to another set. The embedded (inter)relationships of these systems suggests that transitioning to a resilient non-carbonaceous city involves a *whole-system re-conceptualisation* – a creative re-imagining of the future city (Ryan 2013a). This is, in every sense of the word, a *design challenge* – a process that can be considered to have three components. First, envisaging new systems that could support a thriving, culturally satisfying and productive urban future supported by renewable energy. Second, selecting systems from the above that could increase social and physical resilience even as the climate changes. Third, negotiating processes of transition towards those future systems with relevant citizens and other stakeholders.[3] Finally, it has to be expected that no process of creative design, engagement and modelling of future possibilities will be able to achieve more than to suggest plausible new possibilities that *could work*. Given the complexities of socio-cultural-technical-(bio)physical interactions in urban life and the urgency for change-making, potentially the only real hope for successful transformation will emerge from *widespread experimentation* – testing plausible transformed futures *in the real, messy, living world.*

The approach of virtual experimentation, using design research, visioning and engagement, has evolved in response to all the challenges of transforming today's cities. Both projects described in this chapter – Eco-Acupuncture and Visions and Pathways 2040 – take a design approach to research and engagement focused on envisioning 25-year, non-business-as-usual futures for urban neighbourhoods or whole cities. On the basis of that work, they explore policy and design interventions that re-orient the trajectories of current development towards those futures. Great emphasis is placed on the visualisation of possible futures to build support for action necessary to realise them; 'action' encompasses policy, investment in new research innovation and built environments, as well as on-the-ground experimentation (often referred to as living laboratories, see Evans and Karvonen 2014). Through the experience of 6 or more years of this work, it has become clear that the core of the process rests on the interrogation and interpretation of *visualised futures* as the basis for *generating a wide-ranging dialogue with local citizens*. It is this that we characterise as 'virtual city experimentation'.

The methodology of the two projects builds on a field of practice that addresses the negotiation of complex technical and social change often collectively referred to as 'back-casting' – a process that begins with generation of desirable futures and then 'casting' back from those futures to define trajectories of change (Robinson 1988; Dreborg 1996; Quist and Vergragt 2000, 2006; Vergragt and Quist 2011). Such work depends on generating sufficient community 'ownership' of the envisaged futures and participatory visioning has become increasingly accepted as a way to develop such support (Quist and Vergragt 2006; McCormick *et al.* 2013; Ryan 2013b). Participatory visioning is closely related to what is often called 'participatory design', where designers work with clients (often the community) to co-design outcomes. Whilst some interpretations of that process present the designer as a mere facilitator or translator of the 'voice' of the participants, there are other interpretations of the designers' role as more active; the latter is the stance taken in VEIL projects. Manzini (2013), in a critique of the 'passive' approach to participatory design (which he labels 'post-it-note design'), argues for a more active role for designers as 'triggers' to create new conversations and new scenarios of possibilities. There is knowledge and skills that designers can bring to such a process with cultivated openness and dialogic capacity involving the use of images, prototypes, mock-ups, games and so on.

In EcoA and VP2040, the role of designers is tuned to triggering new conversations, to widening the territory to be explored and to interpreting multidisciplinary research (on impacts of climate change, technologies, processes for mitigation and adaptation, and so on) (Gaziulusoy and Ryan 2015). Through this process, the participants' perceptions of 'desirability' and 'plausibility' can be made more transparent, either reaching towards consensual visions or articulating other valid ones. Designers work to visualise 'in-the-mind' reflections on new potentialities, breaking from 'realistic' expectations that assume the continuation of past ('business-as-usual') trajectories. Visualised concepts and images of alternative urban systems are used in an iterative way to generate new conversations about different desirable futures, with feedback from those conversations used to refine propositions. Exhibitions of visualised futures and workshop processes are structured to bring an ever-widening cohort of citizens to review and reflect on alternative futures and pathways. What transpires in that process is virtual experimentation. What has been learnt over time is that the best form of visualisations to engage that in-the-mind interrogation are those that have been defined as *glimpses*: evocations of possible future states that are sufficiently 'open' that they encourage interpretation and translation for the context of the viewer to 'experiment with' rather than a highly defined future that could be interpreted as a blueprint for what *will* unfold (Moy and Ryan 2011).

Case studies

The Eco-Acupuncture (EcoA) programme

EcoA was launched in 2008 as a 'design–research–engagement–action' programme to assist business, communities, towns and cities develop innovations relevant to the decarbonisation of the economy and the development of climate resilient infrastructure. EcoA brought the research capacities of (initially) four universities (all in metropolitan Melbourne) with the post-graduate teaching programmes of the design schools of those universities to the consideration of future challenges and opportunities with a 25-year horizon. The urban precinct and the city became a critical focus for partnerships with local government to:

- examine emerging problems for the future resilience of a specific urban location;
- consider any complex system interactions that form part of those problems;
- visualise future possibilities to resolve identified problems and increase resilience and, most critically;
- design a series of interventions as 'transformation points' towards a resilient low-carbon future.

(Ryan 2013b)

The first experimental investigation of the urban-focused work was in central Melbourne. In contrast to later work, this was not a 'retrofit' of an existing urban precinct but a schematic master-plan vision and framework for a new 'eco-city' on a very large brownfield site close to the central business district. This project gained a great deal of attention, primarily because it was (initially) carried out as an exercise independent of government and private development agencies charged with strategic oversight of the development (who were taking a 'cookie-cutter' approach with little innovation). Media attention focused on the vision/images of the future for the site (VEIL-EBD 2008). From 2009, as the imaginative power of future visions with community engagement became clear, EcoA has developed as a set of partnerships with cities and communities where interest in the opportunities of transformation is high. Over the period 2009–2014 the programme has engaged with five suburbs of metropolitan Melbourne and two country towns in the state of Victoria – Anglesea and Creswick – to explore possibilities for decarbonisation and resilience to very extreme weather events (Ryan *et al.* 2010; VEIL-Broadmeadows 2010; Larsen 2012; Biggs *et al.* 2014). In 2012, an EcoA team travelled to the city of Florence, Italy at the invitation of the city and New York University, Florence for work on the 'Greenaissance' of the UNESCO area of the city (again with resilience and decarbonisation objectives). A similarly sized EcoA project was conducted with the city of Rotterdam in the Netherlands in early 2014, and another project in the city of Leeuwarden in the province of Friesland in the Netherlands commenced in early 2015 with a strong cultural overlay, as this city is the EU Cultural Capital for 2018.

Each EcoA project brings together university researchers, design academics, design masters students and design professionals to work closely for and with local communities. It involves the establishment of highly visible *design ateliers* in a community space. In a sequence of design workshops over a year or more, visions of medium-term (25 years) futures are co-developed, places for near-term interventions are investigated and small-scale, low-cost propositions for those places are designed. The atelier space, regular exhibitions and discussions of visions, are used to build shared ownership of futures and trajectories of development.

The Visions and Pathways 2040 (VP2040) project

VP2040 is a 4-year multi-partner research and engagement project, with funding from the Australian Cooperative Research Centre for Low Carbon Living, to examine the potential for four capital cities in southern Australian states to increase their resilience and reduce their greenhouse gas emissions by 80 per cent by 2040. VP2040 involves a small team of researchers at three universities (University of Melbourne and Swinburne University of Technology in Melbourne Victoria and the University of New South Wales in Sydney) with the collaboration of multinational businesses in design, planning and engineering services and construction, as well as the city councils of Melbourne, Adelaide and Sydney and a number of government departments.

VP2040 builds on the methodological approach of EcoA, projecting visions and scenarios for the future of the cities. It aims to identify intervention points to reach the envisaged futures – in this case for policy and investment for technology, lifestyles, built form and future research. VP2040 works with a loose reference to the framework of the multi-level model of system innovation (Geels 2005) with its three dynamically interacting layers (i.e. landscape, regime, niche). As with the EcoA project, visualised images of future conditions play a significant role in the process – as a way of emphasising the driving forces at the landscape level (not only climate related), projecting possible alternative structures for (future) regimes and implicitly, or explicitly, connecting to niche-level developments that have some emerging relevance. In another framing of the project this is a normative vision process (the re-configuration of transformed systems of provision and patterns of consumption that, together, could decarbonise city life by 80 per cent) with 'back-casting' (Vergragt and Quist 2011) to create narratives and pathways for a trajectory linking the future 2040 conditions to today.

In the 2015 Australian context, the idea of cities reducing their greenhouse gas emissions by 80 per cent in 25 years is a radical deviation from the current political consensus (with a national commitment, as of mid-2015, to a reduction of only 5–8 per cent by 2020 with no targets beyond that date). Thus, it is clear that for any of the future visions or scenarios to be plausible, they have to emerge from some processes of change that are outside the current (mainstream) political imagination. Potential disruptive changes (social and technological) thus form part of the VP2040 research programme and act, in the framing of the project, as niche forces from a multi-level perspective or as a way of structuring the narratives in the back-casting scenario model. Examples of technological disruption include the cost-curve of solar photovoltaic cells, wind power and systems of energy storage, while social innovations include various forms of localisation and voluntary sufficiency movements (Alexander 2014), and mixed socio-technical innovations include social-media platforms for collective consumption (Botsman and Rogers 2014) as well as distributed manufacturing and 3D printing and new organisational models for business and governance (Gaziulusoy and Twomey 2014). All of these involve complex dynamics of change that are sometimes instanced only as niche experiments but sometimes as rapidly growing forces. Envisioning the contribution that these disruptive forces (singularly or in combination) could make to the future is a task that involves speculation, projection and negotiation in the process of co-designing desirable future states (i.e. a process of virtual experimentation).

For this virtual experimentation, the VP2040 project brings together policy makers, professional niche-innovators, activists, designers and researchers in participatory workshops for facilitated future visioning. These workshops aim to get the participants to 'dream' beyond business-as-usual futures[4] and to encapsulate such dreaming in an expanding set of rich visualisations or glimpses of complexly transformed futures that also suggest possible transition narratives and pathways. The dual aim of this process is to expand the understanding of the field of emerging innovations that have the potential to disrupt existing socio-technical regimes (as an input to the research side of the project) *and* to explore, encapsulate and communicate what such futures could plausibly look like as a result of those emerging innovations. Ultimately, communicating these visual encapsulations of ideas becomes one way to widen engagement in the project of experimenting future cities. Figures 5.1 and 5.2 are examples of these future imaginings in the VP2040 project (selected in part to illustrate the range of forms the 'glimpses' may take).

As dialogic objects, the two glimpses are intended to open up possibilities and to challenge expectations about the fixity of the future. Both images are immediately understood by Australian viewers. They show plausible alternative systemic changes within 2040 cities; they evoke transformations that extend beyond the bio-physical qualities of future city life, reflecting social, cultural, economic

Figure 5.1 The transition of a typical Australian street from 2014 to 2040.

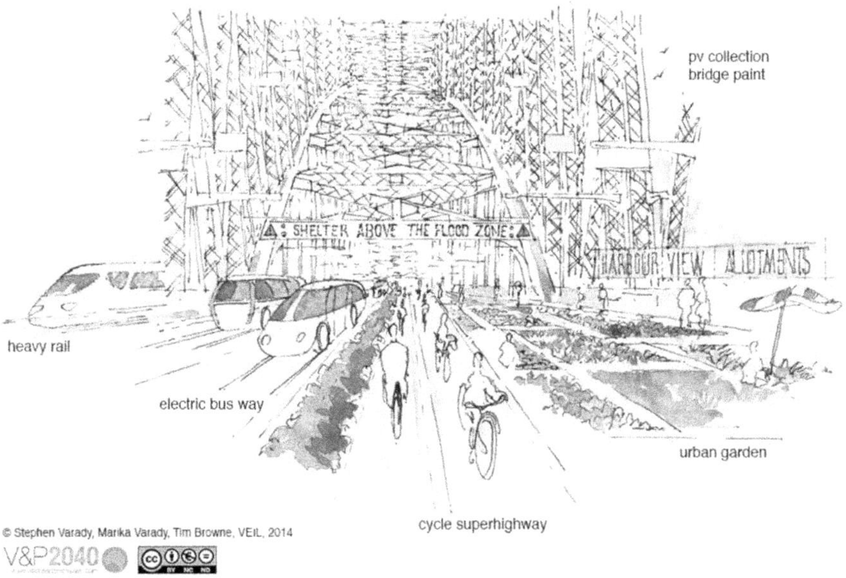

Figure 5.2 The Sydney Harbour Bridge in 2040.

and lifestyle changes. These glimpses are produced by professional designers who attended the visioning workshops. When the glimpses are opened to public gaze, they are accompanied by short interpretive statements that evolve based on dialogue and as the project team constructs more systematic categorisation of future scenarios.[5]

Workshop participants returned to a second workshop session after the glimpses were produced to critically reflect on their plausibility; these glimpses were freely accessible to the public (from the website) and are being used by project partners with their (various) community members. These processes have enabled VP2040 team to identify potential convergences and divergences of ideas about system reconfigurations and value-sets that could affect the trajectories or pathways of change. A widening process of expert consultations is underway to focus in on various aspects of the transformation process.

Learnings and reflections

The process of virtual experimentation through design visioning is presented here as a strategy to address the challenges of rapidly transforming cities, a process to draw-in citizens, researchers, designers, and planners as individuals and as institutions to experiment with different expectations for the future. VEIL has developed a framework to shape the conceptual terrain of that experimental visioning. This includes: optimal time horizons (25 years); the selection of optimistic, desirable changes; the value of distributed systems of provision (more localised, networked, and diverse) for resilience; the representation of trajectories of change as well as new future possibilities (Biggs *et al.* 2010; Ryan 2013b). However, it is a reasonable question to ask of such approaches: can virtual experimentation contribute to the establishment of real-life, on-the-ground, experimentation?

There have been various attempts to track the impact of this process in the longest running EcoA project collaboration. However, tracking the seeding of ideas about the future is complex and problematic. For the longest local government engagement, there have been four EcoA ateliers over a period of 5 years. At a 2011 international conference in Melbourne, that council's senior urban designer cited four broad changes in planning and urban design that, in his view, had been enabled by the process of community deliberations using the EcoA visioning process (Wilson 2011). Two of these interventions in particular – bike paths and community gardens – had seen a reversal of previous community attitudes to those proposed areas of development. At the time of that paper and from follow-up meetings with council officers, there has been growing support for the process as a significant contribution to the council's strategic urban planning and community development projects. That council has committed to a more thorough and open exploration of the impact of such work beginning in 2015.

For the higher profile projects, such as Florence and Rotterdam, assessment of impact is potentially even more complex and difficult because of significant differences in the political, cultural and regulatory underpinnings of the existing regimes. However, both of those projects arose from the desires of the city administration for future thinking that could break from the embedded cultural biases that were seen as limiting the responses to emerging challenges. For Florence, the idea of 'fixity' – that the city is, and must remain, a global treasure, a built museum, preserved from change for global posterity – was recognised by the administration to be in conflict with the growing need to deal with changing climatic conditions. Summer temperatures in mid-tourist season appeared to be regularly reaching around and beyond 40°C and recent winters had brought severe (transport crippling) snowfalls; the city had been experiencing extended drought conditions with altered rainfall patterns that included intense storm events with frequent local flooding. Buildings were being left vacant as they became uninhabitable and Florentines from the rest of the city were reported as surrendering the historic zone to tourism.

Future visions and proposals for intervention provoked strong local debate and even opposition. Two dominant cultural ideas, in particular, were challenged. As a response to the rising heat impacts in the city, the EcoA visions introduced trees in public places such as squares and streets. And in order to reconnect the city inhabitants to its original water source, the Arno, the visions proposed a new physical access to the river via a long floating pontoon-walkway. Both those propositions were seen by some sections of the community as having violated the core of the UNESCO heritage ideals. The Arno access is widely accepted as 'off-limits' to ensure that its fortified walls (that create a physical as well as visual barrier along its path through the UNESCO domain) provide protection against flooding from the river. The EcoA proposals seemed to be the first to challenge the idea of the river as a storm-channel, proposing large up-stream 'sacrificial' wetlands (as a tourist park) and a floating pontoon access to leave the walled fortifications of the banks unchanged. This 'use' of the river was proposed to begin as temporary for spring/summer when water levels are low; its first 'occupation' would be for the revival of the city's lost status as a fashion capital, having the main catwalk for the fashion festival floating on the river just below the Ponte Vecchio.

Throughout the development of those ideas – in the co-creation, visioning process – some local historians pointed to a history that was much more nuanced than the tourist representations of the city's 'renaissance' fabric. Florence has a great history of waves of innovation that included transformation of its urban conditions. There is a wealth of images and stories about the role of the Arno in the making of the city – for waste disposal for industry and for river transport of goods. Those images also illustrate the progressive loss of vegetation that once spread outwards from the Arno banks to feed the city. The large, open, stone-paved piazze that seem now to characterise the fabric of the old city were actually remodelled in the mid-nineteenth century to reflect the city's status as Italy's capital, embracing new ideas of metropolitan planning. This included the removal of their markets and the widening of some of the central streets of the city.

Another prominent representation of the city's era as capital was its grand cast-iron central market designed and constructed in the 1870s. This building had fallen from use as locals shunned the UNESCO area, frequenting other food markets instead, leaving a strange and often problematic tourist facility that operated for limited hours each week. The most successful of the EcoA re-imaginings of the city involved proposals for the re-functioning of that building and its square, to re-establish a site within the old town that was truly Florentine. The evocative visual images for that revitalisation were quickly taken up and became the basis for a tender for the redevelopment of the facility – that was completed in 2014. The system connections that were proposed for that redevelopment in the EcoA visions have to date had no impact. The new tram network that stalled after a prolonged and difficult completion of the first of five proposed lines was (re)visioned as a mixed mode transport link, amongst other things carrying organic waste from the city centre during the night.

Based on the evaluation of 4 years of EcoA design visioning, there appears to be strong support for glimpses that generate curiosity (often seen as amusing as well as challenging) and glimpses that draw the viewer into engaging with *ideas*, requiring them to *transpose the re-aligned conditions depicted, into their own street, neighbourhood or daily life*. However, there is the need for ongoing experimentation about the optimum form for visions that can successfully stimulate virtual city experimentation. The effectiveness of visioning resides in achieving a balance between adequate representation, or the 'fidelity' of a glimpse or vision, and some level of ambiguity or lack of resolution to invite a dialogic response. Visions must act as a provocation for a conversation where the viewer feels curiosity as well as comfort in interpreting the image rather than merely being subject to its authority. However, too little detail or resolution in the glimpse or vision can lead to confusion and a lack of engagement. This balance is not set or universal. Different audiences and viewing contexts require different levels of resolution. Adjusting this balance between openness and fidelity, to provide a sufficient framing of new system possibilities is part of the ongoing collaboration with some of the participating Melbourne councils. This will extend to testing some of those representation skills of designers beyond the two-dimensional format for glimpses; 'story-boarding' or sketches of daily life and spatial changes over time have regularly been incorporated in glimpses, testing of three-dimensional representations of futures may well be the next step.

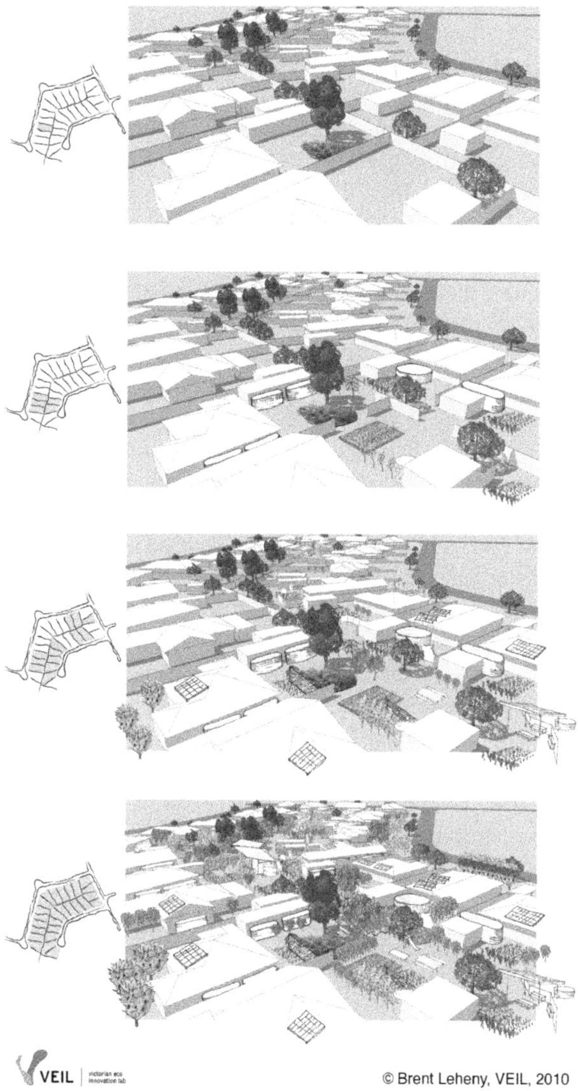

Figure 5.3 Change over time in the transformation of suburban space and the 'collapse' of private backyards.

Conclusions

The virtual experimentation approach discussed in this chapter defines three categories of glimpses that are essential to the process. First, glimpses for inspiration that aim to shift the boundaries of what is considered permissible, desirable and possible for future conditions and systems (Ryan 2013b). Second, glimpses of (possible) trajectories of change that open up a discussion of how a non-development-as-usual future might unfold. Third, glimpses of niche interventions that are intended to build a constituency for their realisation soon. The overarching design objective for these is that they evoke possibilities for new system architectures (such as distributed water, food, energy, transport and shelter).

For EcoA and VP2040, local councils and community representatives required that these should be low cost and low risk in the event of failure. This suggests an important new parameter for 'living laboratories' and real-life experimentation.

Visioning transformative futures is a difficult process, more so for people without any experience of systematic scenario building. One challenge that frequently arises is overcoming deeply ingrained perceptions about the future and the nature of change processes. From the experience of EcoA and the VP2040 projects, it is apparent that the public conception of the future reflects their perception of how the world works and how and at what *rate* change *can take place*. The VEIL visioning processes frequently reveal the tendency of participants to project the status quo (development-as-usual) well beyond its feasible lifespan. In the absence of processes that assist participants to become sensitive to disruptive forces in the flow of change, people's conceptions of the future will tend to involve only mild deviations from the trajectories of current development. Enabling participants to conceive non-linear depictions of the future requires an immersive, layered, and iterative journey. Design visioning helps to catalyse a cognitive break from the present.

Virtual experimentation achieves two important outcomes relevant to enabling radical transformation of city systems. Glimpses can make explicit what changes a particular group of people (the vision creators) deem possible and how they are linked to present conditions and understandings of future challenges. This latter point is critical because it can help identify – or reveal – aspects of the present that could act as launch pads for routes to a transformed world (a terrain of disruptive potential). Also, by making explicit a set of assumptions of the future, transformative visions can become an effective means for cultivating a dialogue about the ways in which participants (and essentially 'system innovators') understand the possibilities of change. Virtual experimentation can create experiences of the future for developers and opportunities and space for real physical experimentation.

Notes

1 Marx (1867) is often quoted to this effect when he sought to distinguish between an architect and a bee – both create extraordinary structures but only humans can endlessly vary the design of the structures in their imagination – and consider their relative advantages and disadvantages – before committing to action. '*A bee puts to shame many an architect in the construction of her cells. But what distinguishes the worst of architects*

from the best of bees is this, that the architect raises his structure in his imagination before he erects it in reality'.

2 See for example the Biophilic Cities movement: http://biophiliccities.org [October 2015].

3 This last step can present a significant hurdle. Beck (2010) has pointed to the urgent and critical problem for the greening of society – gaining everyday support 'from below', support for transformations that can appear to undermine current lifestyles, consumption habits and practices and established systems of social status. In the process of negotiating a viable future, the challenge is to embrace forms of engagement for current urban citizens that could give some hope that such futures would be seen as desirable.

4 See: http://www.visionsandpathways.com/about/vp2040-video/ [October 2015].

5 For example, see: http://www.visionsandpathways.com/research/visions/ [October 2015].

References

Alexander, S. (2014). *Disruptive Social Innovation for a Low-carbon World.* Melbourne: Victorian Eco-innovation Lab.

Beatley, T. (2011). *Biophilic Cities. Integrating Nature into Urban Design and Planning.* Washington, DC: Island Press.

Beatley, T. and Newman, P. (2013). Biophilic cities are sustainable, resilient cities. *Sustainability*, 5: 3328–3345.

Beck, U. (2010). Climate for change, or how to create a green modernity? *Theory, Culture and Society*, 27: 254–266.

Bettencourt, L.M. and West, G.B. (2011). Bigger cities do more with less. *Scientific American*, 305(3): 52–53.

Bettencourt, L.M., Lobo, J., Helbring, D., Kühnert, C. and West, G.B. (2007). Growth, innovation, scaling, and the pace of life in cities. *Proceedings of the National Academy of Sciences of the United States of America*, 104: 7301–7306.

Biggs, C., Ryan, C., and Wiseman, J. (2010). *Distributed Systems: A Design Approach for Sustainable and Resilient Critical Infrastructure.* Melbourne: Victorian Eco-Innovation Lab.

Biggs, C., Ryan, C., Bird, J., Trudgeon, M. and Roggema, R. (2014). *Visions of Resilience: Design-led Transformation for Climate Extremes*, Victorian Eco-Innovation Lab, University of Melbourne. [Online]. Available: http://www.ecoinnovationlab.com/visions-of-resilience [25 June 2014].

Botsman, R. and Rogers, R. (2010). *What's Mine is Yours: The Rise of Collaborative Consumption.* New York: HarperCollins.

Dreborg, K.H. (1996). Essence of backcasting. *Futures*, 28: 813–828.

Evans, J. and Karvonen, A. (2014). 'Give me a laboratory and I will lower your carbon footprint!' – Urban laboratories and the governance of low-carbon futures. *International Journal of Urban and Regional Research*, 38: 413–430.

Gaziulusoy, A.I. and Ryan, C. (2015). Low-carbon, resilient, city futures – a design-mediated approach: visions and pathways 2040. Paper presented at the *8th Making Cities Liveable Conference*. Melbourne, 6–7 July 2015.

Gaziulusoy, A.I. and Twomey, P. (2014). *Emerging Approaches in Business Model Innovation Relevant to Sustainability and Low-carbon Transitions.* Melbourne: Victorian Eco-innovation Lab.

Geels, F.W. (2005). *Technological Transitions and System Innovations: A Co-evolutionary and Socio-technical Analysis.* Cheltenham: Edward Elgar.

Glaeser, E. (2011). T*riumph of the City: How Our Greatest Invention Makes Us Richer, Smarter, Greener, Healthier and Happier.* New York: Penguin Press.

Grimm, N.B., Faeth, S.H., Golubiewski, N.E., Redman, C.L., Jianguo, W., Bai, X. and Briggs, J.M. (2008). Global change and the ecology of cities. *Science*, 319: 756–760.

Hajer, M. (1995). *The Politics of Environmental Discourse: Ecological Modernization and the Policy Process*. Oxford: Clarendon Press.

Hajer, M. and Dassen, T. (2015). *Smart About Cities: Visualising the Challenge for 21st Century Urbanism*. Rotterdam: NAI Publishers.

Harvey, D. (2012). *Rebel Cities: From the Right to the City to the Urban Revolution*. London: Verso.

Hughes, L. and Steffen, W. (2013). *The Critical Decade 2013: Climate Change Science, Risks and Response*. Canberra: Climate Commission.

Kareiva, P. and Marvier, M. (2007). Conservation for the people. *Scientific American*, 297: 50–57.

Katz, B. and Bradley, J. (2013). *The Metropolitan Revolution: How Cities and Metros are Fixing Our Broken Politics and Fragile Economy*. Washington, DC: Brookings Institution Press.

Kautto, N. and Ryan, C. (2015). *Cities as Agents of Disruptive Change – Reflections on Global Cities with Significant Decarbonising Strategies and Programs*. Visions and Pathways 2040. Foreground paper. [Online]. Available: www.visionsandpathways.com [1 October 2015].

Landes, D.S. (1972). *The Unbound Prometheus: Technological Change and Industrial Development in Western Europe from 1750*. Cambridge: Cambridge University Press.

Larsen, K. (2012). *Can Food Hubs Catalyse Healthy and Resilient Local Food Systems in Victoria?* Victorian Eco-Innovation Lab, University of Melbourne. [Online]. Available: http://www.ecoinnovationlab.com/research/current-research/520-food-hubs-in-victoria [21 June 2013].

McCormick, K., Anderberg, S., Coenen, L. and Neij, L. (2013). Advancing sustainable urban transformation. *Journal of Cleaner Production*, 50: 1–11.

Manzini, E. (2013). Against post-it design: to make things happen. [Online] Available: http://www.desis-network.org/forums/against-post-it-design-make-things-happen. [13 May 2015].

Marcus, L. and Colding, J. 2011. Towards a spatial morphology of urban social-ecological systems. Paper presented at *18th International Conference on Urban Form*. Concordia University, Montreal, 26–29 August 2011.

Marx, K. (1867). *Das Kapital*. Volume 1. Moscow: Progress Publishers.

Moy, D. and Ryan, C. (2011). Using scenarios to explore system change: VEIL, Local Food Depot. In Meroni, A. and Sangiorgi, D. (eds), *Design for Services*. London: Gower, 161–171.

Quist, J. and Vergragt, P.J. (2000). System innovations towards sustainability using stakeholder workshops and scenarios. Paper presented at the POSTI Conference on *Policy Agendas for Sustainable Technological Innovation*. London, 1–3 December 2000.

Quist, J. and Vergragt, P. (2006). Past and future of backcasting: the shift to stakeholder participation and a proposal for a methodological framework. *Futures*, 38: 1027–1045.

Robinson, J.B. (1988). Unlearning and backcasting: rethinking some of the questions we ask about the future. *Technological Forecasting and Social Change*, 33: 325–338.

Ryan C. (1985). The nature of construction is the re-construction of nature: nature, technology and ideology in the design of the built environment. In Dovey, K., Downton, P. and Missingham, G. (eds), *Place and Placemaking Conference: The Third International Conference of the Association for People and the Physical Environment Research*. Melbourne: PAPER books.

Ryan, C. (2013a). *VEIL and Urban 'Eco-Transformation'*. Victorian Eco-Innovation Lab, University of Melbourne. [Online]. Available: http://www.ecoinnovationlab.com/project_content/veil-urban-eco-transformation/ [25 June 2014].

Ryan, C. (2013b). Eco-Acupuncture: designing and facilitating pathways for urban transformation, for a resilient low-carbon future. *Journal of Cleaner Production*, 50: 1–11.

Ryan, C., Trudgeon, M., Moy, D., Larsen, K., Biggs, C., Archdeacon, A. and Eales, R. (2010) *Victorian Eco-Innovation Lab Annual Report*. [Online]. Available: http://www.ecoinnovationlab.com/content/publications/ [25 June 2015].

Satterthwaite, D. and Dodman, D. (2009). The role of cities in climate change. In Engelman, R., Renner, M. and Sawin, J. (eds), *State of the World 2009: Into a Warming World*. Washington, DC: Worldwatch Institute, 75–77.

Stone, B. (2012). *The City and the Coming Climate*. Cambridge: Cambridge University Press.

UNEP (2011). *Cities: Investing in Energy and Resource Efficiency*. Paris: United Nations Environment Programme.

VEIL-Broadmeadows (2010). Vision Broadmeadows 2032. *Eco-Acupuncture Enabling Localised Design Interventions*. Victorian Eco-Innovation Lab, University of Melbourne. [Online]. Available: http://www.ecoinnovationlab.com/content/publications/ [16 May 2014].

VEIL–EBD (2008). *EBD: Ecological Business District. Zero Carbon Eco city*. Victorian Eco-Innovation Lab, University of Melbourne. [Online]. Available: http://www.ecoinnovationlab.com/content/publications/ [15 April 2014].

Vergragt, P.J. and Quist, J. (2011). Backcasting for sustainability: introduction to the special issue. *Technological Forecasting and Social Change*, 78: 747–755.

Wilson, M. (2011). Case study: Vision: Broadmeadows 2032. Paper presented to the *International Thriving Neighbourhoods Conference*. Melbourne, 2011. [Online]. Available: http://www.thrivingneighbourhoods.org/presentations/2011-presentations/ [14 October 2012].

6 The boundaries of experimentation in sustainable urbanism

Elizabeth Rapoport

Introduction

The disciplines of urban planning and design have long directed themselves towards the achievement of particular objectives, from improving public health to facilitating efficient movement (Hall 2002). More recently, as Gunder and Hillier (2009) have argued, in many countries achieving 'sustainable development' has become a driving objective for planners. There are many different types of planning interventions that aim to achieve sustainability. These range from policies that incentivise particular behaviours and technologies, to community-led initiatives such as Transition Towns, to large infrastructure projects and new eco-cities.

This chapter focuses on one particular type of sustainability oriented urban planning intervention: large-scale sustainable urban projects (SUPs). SUPs have several key characteristics. They are efforts to create a new piece of city, often from scratch. Their size and scale ranges from a large property development to an entire new town. They are mixed-use; that is, they incorporate multiple land uses (e.g. residential and commercial). They are also urban in that they are either located in an existing urban area, or aim to create a new one. And finally, they are sustainable in that they explicitly attempt to integrate sustainable design and planning interventions.

Projects of this type, which are sometimes called eco-cities, have proliferated in recent years (Joss 2010). A spate of eco-city projects in China in recent years include the now-moribund Dongtan Eco-City, designed in 2005 by the British engineering firm Arup for a site near Shanghai, and the Tangshan Caofeidian International Eco-City, planned with the input of the Swedish engineering firm Sweco and currently under construction (Joss and Molella 2013). This trend is not limited to China. In India, along the new expressway between Mumbai and Pune, the privately developed sustainable city of Lavasa, master-planned by the American architecture firm HOK, is currently emerging (Datta 2012). In the desert outside of Abu Dhabi, the low carbon Masdar City, master-planned by the British architecture firm Foster and Partners, is also taking shape.

Such projects usually claim to apply a tripartite (environmental, economic, social) definition of sustainability. They incorporate a range of sustainable urban planning and design principles and technologies designed to increase resource efficiency, lower operating costs and deliver a high quality of life for residents.

These often include passive design – building and urban design approaches where building height, massing and orientation are designed in such a way that they reduce the need for artificial heating, cooling and ventilation. They may also incorporate alternative approaches to urban infrastructure, such as sustainable urban drainage systems that integrate surface water retention and drainage into landscape design. Projects often also include more active technologies, such as district heating or cooling, in which heating and cooling is provided from a central plant and distributed via an underground network of pipes to feed building systems. Various types of low carbon public transport systems, such as bus rapid transit, light rail and trams are also popular.

SUPs, then, are experimental in their attempts to apply, at scale, ideas that aim to increase the sustainability of urban design, planning and infrastructure. Yet they are usually planned and delivered through a conventional approach to urban development, driven by a large developer and required to deliver on economic as well as environmental objectives. Given this, I argue that SUPs are also experimental in a second way. These projects experiment with the management and governance of urban development, testing whether sustainability objectives can be incorporated into an entrepreneurial mode of building new urban places. In this chapter, I focus primarily on this latter form of experimentation and explore how it shapes the former.

In the chapter, I unpack some of the critical factors driving the incorporation of sustainability objectives into conventional urban development, and demonstrate how these drivers shape the type and extent of planning and design experimentation that occurs. The chapter draws on research conducted between 2010 and 2012 on the international industry of built environment firms working in the field of sustainable urban planning and design. Research included over 50 interviews and extended participant observation with architects and planners working at prominent international design firms as well as their developer clients, and content analysis of the plans for a number of projects. In the next section of this chapter, I consider the status of sustainable urban projects as experiments. I then trace the drivers behind their embrace of sustainability, linking this to the growing economic value that sustainability has in property development and marketing. The nature of these drivers of experimentation, I conclude, results in SUPs taking an incremental rather than comprehensive approach to applying sustainable urban planning and design principles at scale.

Experiments in sustainable urbanism

What qualifies sustainable urban projects as 'experiments?' Bulkeley and Castán Broto (2013: 363), in discussing urban climate change experiments, define these as 'interventions in which there is a more or less explicit attempt to innovate, learn or gain experience'. Karvonen and van Heur (2014: 383) define experimentation as: '(1) involving a specific set-up of instruments and people that (2) aims for the controlled inducement of changes and (3) the measurement of these changes' as well as involving 'a double move of observation and intervention'. Building on these definitions, urban experiments can be conceived as interventions in urban systems that aim to produce and test knowledge. They seek to provide evidence that can be used to inform future interventions.

For urban planners, such evidence can be a valuable resource. The extent to which particular planning and urban design interventions can actually be associated with achieving sustainability objectives is heavily contested (Bulkeley and Betsill 2005; Williams 2009). This is in part because urban planning and policy interventions are usually layered on an existing city with all its complexity, making the tasks of observation, intervention and monitoring challenging. SUPs are distinctive in two ways that make them particularly valuable as urban experiments. First, they often start from a tabula rasa, constructed either on a vacant site adjacent to an existing city, or on a cleared site within a city. This makes it easier, first of all, to incorporate sustainable design ideas and technologies that are difficult to retrofit into an existing urban fabric, as well as those that are most effective at scale. Starting from scratch also makes it easier to observe and measure the impact of these interventions.

A second way that SUPs are distinctive is that they test more than a single technology or design idea. Smaller projects can produce and test knowledge about the impact of a particular sustainability intervention. For example, can a new approach to distributing energy or managing surface water runoff deliver the returns that models and calculations predict? Does providing ample public transit lead to reduced automobile emissions? In theory, SUPs can go beyond this to test the impacts of an integrated and holistic approach to sustainable urban planning and design. They are experiments in sustainable urbanism at scale, opportunities to envision what a sustainable city would look like if it could be built from the ground up.

Like most urban experiments, SUPs are created to do more than just produce and test knowledge. A low carbon building provides housing and office space, rainwater harvesting technology provides water, and a light rail system moves people around. Perhaps most importantly in the context of this chapter, these projects are property developments that are usually expected to deliver a return on investment. Understanding the expectations that the developers of SUPs have for their projects, and how these expectations shape experimentation, is critical. Unpacking them can help explain how and why urban experiments can be oriented towards broader strategic objectives that may or may not relate to their ostensible experimental purpose (Evans and Karvonen 2014).

Experiments, particularly those produced through networked governance arrangements (often in the form of public–private partnerships) can serve to reinforce and consolidate dominant agendas and power relations (Evans and Karvonen 2014). To this end, identifying who is experimenting is essential (Evans 2011; Bulkeley and Castán Broto 2013). Just as important is to understand why they are doing so. The remainder of this chapter explores the drivers to incorporate sustainable urbanism principles into conventional property development projects, and the way that these drivers shape the type of experimentation that occurs.

Sustainable property development: drivers of experimentation

Incorporating sustainable design principles into urban projects is not new, but historically it has occurred in small-scale projects driven by communities, local governments and not-for- profit organisations (Rapoport 2014). The integration of sustainability into larger, more commercially oriented projects began in the first decade of the twenty-first century. This occurred in a context of the late twentieth century shift to more entrepreneurial forms of urban governance. In this approach, the role of the state in planning in many countries around the world is restricted largely to the commissioning and oversight of urban planning and development services (Harvey 1989; Hall and Hubbard 1996). Direct state subsidies for urban development are often minimal. This means that large-scale urban planning projects are effectively conventional property developments and must deliver a financial return within a specific time frame (e.g. 20 years). In an entrepreneurial governance approach, urban development is carried out using a partnership model of financing and delivery that requires harnessing the support of a wide variety of actors (Bassett 1996; Kjær 2004; Shaw 2013). To achieve this, governments often outsource the conception and management of large-scale urban development projects, or simply sell off or lease the land to be developed. Thus the developers behind SUPs tend to be either quasi-governmental but largely autonomous development corporations, or private developers.

This entrepreneurial climate introduces challenges for incorporating sustainability features into an urban project. Urban development has effectively become property development, and property development is, by and large, a profit-oriented, risk-adverse industry. Bringing sustainability features into an urban project requires motivating a wide range of actors to agree to depart from their established ways of doing things. This section reviews some of the broader factors influencing the property development industry that have encouraged developers to embrace sustainability.

The growing importance for businesses to make a public and visible commitment to 'Corporate Responsibility' (CR) is a critical driver behind the incorporation of sustainability into commercial development projects. According to research carried out in 2013 by the international accountancy firm KPMG, regular CR reporting is standard business practice internationally, with nearly three-quarters of 4,100 companies surveyed producing a CR strategy (KPMG 2013). As the importance of CR has grown, a number of voluntary international standards have emerged that companies can sign up to in order to demonstrate their commitment to responsible business and investment practices. A number of these voluntary standards focus on sustainability. One example is the Global Reporting Initiative (GRI), an organisation whose mission is to encourage organisations to produce regular sustainability reports in the same way that they produce financial reports. GRI provides a set of guidelines to support organisations to create a sustainability report. Organisations can choose to make their reports public through GRI's online sustainability disclosure database. In May 2015, over 7,500 organisations around the world had done so (GRI 2015). According to the KPMG report cited above, 78 per cent of surveyed companies use GRI guidelines when preparing their CR or sustainability strategies.

Companies can use their CR programmes and sustainability strategies to demonstrate that they are a sustainable and responsible company to invest in. Being able to demonstrate this has become increasingly important, as a large number of the world's biggest investors are signatories to a voluntary scheme encouraging responsible investment, the Principles for Responsible Investment initiative (PRI), launched in 2006 and supported by the United Nations. Signatories publicly commit to adopting and implementing six principles for environmental, social and corporate governance. PRI now counts over 1,300 signatories who collectively manage over US$59 trillion in assets (PRI 2015). Current signatories include 910 investment managers including many of the world's largest pension funds.

In an interview, an expert in financing sustainable urban development projects explained that this scheme is 'having all kinds of impacts on companies because companies now invest in ways that are in alignment with principles for responsible investing. Or else they can't get the institutional capital that makes their businesses go.' For property developers, this means that access to investment capital may require a demonstrable commitment to sustainability. For example, pension funds, many of which are signed up to PRI, are currently increasing their levels of investment in property and infrastructure (Flood 2013). In addition, to comply with their own CR commitments, the companies that buy or lease space in an urban development increasingly require that the development demonstrates a commitment to sustainability.

This demand that property developers face to demonstrate that their projects are sustainable has led to the increased adoption of another type of voluntary standard: building and urban development sustainability rating and certification systems. The two most popular systems internationally are the US Green Building Council's Leadership in Energy and Environmental Design (LEED) and the UK's Building Research Establishment Environmental Assessment Methodology (BREEAM) (Saunders 2008). Developers that want their project to be LEED- or BREEAM-certified must accumulate points or credits by incorporating sustainability features into their design.

These tools serve a dual purpose, providing guidance to designers on how to make a development project sustainable, and a certification that it has achieved this objective. A LEED-certified project, for example, can use the LEED logo on promotional material and place a plaque on the building itself. For corporate tenants, occupying a certified building is a way of making good on commitments in their CR and sustainability strategies. The website for the multinational bank HSBC, for instance, includes a commitment to achieving LEED certification for their top 50 energy-consuming buildings (HSBC 2014). HSBC is just one of many companies that have made the decision to publicly commit to a preference for LEED-certified office space.

Having such a certification was originally a way to differentiate a project, but today is increasingly a requirement for high-end property developments that seek to attract international corporate tenants. A US-based sustainability engineer explained the situation: 'Part of doing Class A office building in New York or San Francisco is you have to have LEED Gold because everyone else does and if you don't, you're a loser.' The director of a large urban regeneration project in Germany said that developers' ambitions to attract corporate tenants, who in turn demand sustainability, mean that they have no choice but to meet these demands. A Hong Kong based property developer described how just over a decade ago developers in that city dismissed the idea of such certifications. In his view, once one prominent development, the Hong Kong International Financial Centre (IFC), used its LEED accreditation in its marketing campaign, other developers then realised they had to catch up with this 'international standard' (see Table 6.1).

As these quotes demonstrate, the property developers behind SUPs are motivated to experiment with sustainable urbanism in order to remain competitive in their quest for investment and high-value tenants. Sustainability, then, is incorporated into urban projects not for its intrinsic value, but for its instrumental value in helping make a project viable. However, as I will discuss in the next section, justifying sustainable urbanism largely on the basis of its economic merits limits the scale and scope of the experimentation that occurs.

Table 6.1 Summary of systems and standards driving the uptake of sustainability principles in urban development

System	*Creator*	*Established*	*Purpose*	*Usage*
Principles for Responsible Investment initiative (PRI)	United Nations	2006	Voluntary scheme encouraging responsible investment.	There are three categories of signatories: asset owners, investment managers and service providers.
Global Reporting Initiative (GRI) sustainability reporting guidelines	Global Reporting Initiative	1999	Voluntary guidelines to support organisations to prepare their sustainability reports.	Used by organisations (largely businesses) in countries around the world.
Leadership in Energy and Environmental Design (LEED)	United States Green Building Council	2000	Environmental assessment method for buildings and urban developments.	Used by property developers on building and urban-scale projects.

Grand designs, incremental experiments

The previous section demonstrated how sustainability's perceived value in achieving economic objectives, in particular helping a project compete for investment and tenants, is a primary driver for its incorporation into large urban projects. SUPs are meant to be ambitious experiments in sustainable urbanism at scale, testing out new designs and technologies. Dongtan was billed as 'a global template for sustainability in urban planning' (Bullivant 2007: 127), while the original aim of Masdar City was to be zero-carbon. In this section, I explore how the drivers of experimentation identified above influence, and ultimately limit, the type and range of sustainable designs and technologies that SUPs experiment with.

Most property developers are funded by publicly listed investors who look for a return on their investment in a specific time frame. This naturally makes them adverse to any risks that might threaten their ability to do so. This risk aversion is an incentive to use proven and established designs and designers over more untested or experimental options. In practice, this means that the scope and range of sustainable designs and technologies that may be incorporated into a SUP will be limited, as will the range of companies who are considered to design the project.

Developers are familiar with the rate of return and potential risks of investing in conventional infrastructure or housing stock, less so with those for putting in a light rail system or a combined heat and power plant. In addition, if they do not plan on retaining ownership of the site, some of the longer-term financial benefits of alternative and more sustainable approaches are not an incentive. For example, the lower operating costs of district energy and longer-term benefits of a reliable and guaranteed energy supply are of limited interest to an entity that is planning to sell a development fairly quickly.

The benefits that sustainability brings in terms of branding and marketing and securing investment offsets risks and increased costs, but may offer diminishing returns. It may be possible to convince a property developer of the value of obtaining LEED Platinum certification (the highest level awarded) but it will be difficult to demonstrate the commercial value in going beyond this standard, or applying new and alternative approaches that perhaps do not lead to achieving a credit in the LEED system. One US-based architect explained the situation by creating an analogy with the Tour de France:

> If you think of it as the Tour De France, they (developers) want to be at the front of the peloton but they don't want to be in the little breakaway group that's five miles ahead of the peloton because if they are in that little breakaway group, it means they have over-committed themselves financially.
>
> (Paul Schlapobersky, Senior Associate, Machado and Silvetti Associates, Boston, USA)

Ultimately, the developers of SUPs want to be competitive but few are interested in being at the cutting edge. This also goes some way to explaining the popularity of systems like LEED. When urban sustainability becomes part of a business strategy, it needs to be able to be monitored, evaluated and reported on.

This requirement leads to the popularity of an approach that is standardised, recognisable and marketable.

LEED, BREEAM and other recognised certifications can be helpful for property marketing purposes. Even more valuable is to be able to combine this with the name of a recognisable international design firm, as this can help attract the investment and government support needed to get a project off the ground (Olds 2001; Wu 2007). Most SUPs are high-profile projects, with the designers selected through invited competitions open to a limited number of international firms. As a result, many SUPs are planned and designed by a relatively small group of private sector architecture, engineering and planning consultants (Rapoport 2015).

This situation has led to the development of what several interviewees referred to as a list of 'usual suspects' of a small number of firms that are repeatedly hired to work on SUPs. In some ways, this supports the international uptake of sustainability in large urban projects, as these companies stake their reputation on being cutting edge and innovative. Most have dedicated sustainability teams that are tasked with continuously improving their company's application of sustainability principles to urban projects. However, ultimately SUPs are experiments designed by a small group of people, most of whom were educated in what one interviewee described as a 'European model of urbanism'. Their descriptions of sustainable urbanism are remarkably consistent as are the sustainability ideas presented in master plans.

Learning from sustainable urban projects

If SUPs are experiments in the management and governance of urban development, testing whether sustainability objectives can be incorporated into today's dominant entrepreneurial property development mode of building new urban places, then what have we learned from them so far? Clearly more evidence is needed to answer this question, but I would like to put forward a number of preliminary proposals.

First, the experimentation carried out in SUPs of incorporating sustainability into property development does have a positive and tangible impact on the industry. As the interviewees highlighted, making a commitment to sustainability is important in some segments of the property development industry. This is evidenced by the number of developers who choose to have their commitment certified. A total of 388 projects around the world have been certified as achieving LEED's standard for urban-scale projects, LEED for Neighbourhood Development, since it was introduced in 2010 (USGBC 2014). High-profile projects attract substantial attention from professionals and academics working in the built environment, as well as the general public. When they publically embrace sustainability, this contributes to keeping the sustainability of the built environment on the public agenda. Meaningful and impactful innovation in planning and design, however, requires new and innovative models of financing and delivering large urban projects. This is being addressed by a small group of bespoke investment banks who are working to convince investors to look at longer-term time horizons that make investment in sustainable design and infrastructure feasible.

Second, while it is possible to redirect urban development in a more sustainable direction within the confines of conventional urban planning paradigms, the shift will be more incremental than transformational. As discussed above, developers do not want to expose themselves to too much risk. This conclusion, supported by evidence from a growing body of research on individual projects, demonstrates that SUPs have yet to provide examples of transformative change (Keeton 2011; Datta 2012; Chang and Sheppard 2013; Cugurullo 2015). In addition, the economic drivers of sustainability's incorporation into SUPs mean that 'sustainability' in the urban realm comes to be equated with interpretations of the concept that can be easily recognised and marketed, such as LEED.

This draws attention to the fact that it is important to critically consider the type of sustainability that SUPs experiment with. Building on this point, my third observation is that SUPs are experimenting with an economically oriented approach to sustainability. In SUPs, sustainability is used instrumentally to support the achievement of traditional objectives of attracting tenants and investment, rather than being valued in its own right. A number of interviewees told me that if sustainability features do not 'stack up financially' on a project they often end up being cut out of the plan. More concerning, and resonant with the findings presented here, a number of researchers have raised concerns that eco-city projects emphasise economic development over environmental and social issues (Datta 2012; Caprotti 2015; Cugurullo 2015). Broader critiques focus on the potential for such projects to create ecological enclaves, and the appropriation of land and exploitation of workers that often go into their making (Hodson and Marvin 2010; Caprotti 2014).

Conclusion

This chapter has linked the experimentation in sustainable urbanism seen on large-scale sustainable urban projects to trends outside of the urban development industry. These include the increasing numbers of investors signing up to PRI and the popularity of corporate responsibility strategies and sustainability reporting. These trends are making it more important for high-end property developments to incorporate a visible commitment to sustainability to attract the resources needed to get a project off the ground. Sustainability, then, has come to have economic value. To capitalise on this value, developers are experimenting with changes to conventional approaches to delivering urban projects. This experimentation consists of testing new models of ownership and revenue generation as much as particular designs and technologies.

Many SUPs make grand claims to be models of a sustainable urban future. However, what they ultimately deliver is an incremental approach to incorporating sustainability principles into urban development projects in which economic concerns remain paramount in the interpretation of sustainability. In addition, linking the prospects for innovation in sustainable urbanism to its economic value means that sustainability's continued popularity depends upon its commercial value. If the utility of an approach is tied largely to its marketing value, what happens if this fades? Today, sustainability is now so commonly stated as an objective for large urban projects that it may no longer be a marketable differentiator. Recently, newer buzzwords such as resilient and smart have replaced sustainability at the forefront of the built environment industry's attention.

The case of SUPs also tells us something about urban experiments more broadly. The factors that make experimentation possible can also limit the scale and scope of experimentation that can occur. The competitive nature of the international property development industry has opened the door to experimentation with sustainable urbanism. Yet the risk-averse nature of this industry means that it favours limited, incremental experiments.

SUPs may not be the ideal place to experiment with sustainable urbanism. High-profile urban projects such as eco- or smart-cities are often used to support broader agendas, both political (Cugurullo 2015; Datta 2015) and commercial. Sustainability will ultimately be subservient to such agendas. Smaller-scale, community, civil society or local-government driven initiatives present alternative models of experimentation, driven by different motivations, and may produce lessons that can be scaled up to larger projects. We should be cautious about pinning our hopes for sustainable urban futures on an approach that is still only tinkering around the edges of conventional urban development.

References

Bassett, K. (1996). Partnerships, business elites and urban politics: new forms of governance in an English city? *Urban Studies*, 33: 539–555.

Bulkeley, H. and Betsill, M. (2005). Rethinking sustainable cities: multilevel governance and the 'urban' politics of climate change. *Environmental Politics*, 14: 42–63.

Bulkeley, H. and Castán Broto, V. (2013). Government by experiment? Global cities and the governing of climate change. *Transactions of the Institute of British Geographers*, 38: 361–375.

Bullivant, L. (2007). Dongtan – a Shangri-la for Shanghai? *Architecture and Urbanism*, 440: 122–127.

Caprotti, F. (2014). Eco-urbanism and the eco-city, or, denying the right to the city? *Antipode*, 46: 1285–1303.

Caprotti, F. (2015). *Eco-Cities and the Transition to Low Carbon Economies*. London: Palgrave Macmillan.

Chang, I.-C.C. and Sheppard, E. (2013). China's eco-cities as variegated urban sustainability: Dongtan Eco-City and Chongming Eco-Island. *Journal of Urban Technology*, 20: 57–75.

Cugurullo, F. (2015). Urban eco-modernisation and the policy context of new eco-city projects: where Masdar City fails and why. *Urban Studies*, earlyview online.

Datta, A. (2012). India's ecocity? Environment, urbanisation, and mobility in the making of Lavasa. *Environment and Planning C: Government and Policy*, 30: 982–996.

Datta, A. (2015). New urban utopias of postcolonial India 'entrepreneurial urbanization' in Dholera smart city, Gujarat. *Dialogues in Human Geography*, 5: 3–22.

Evans, J. (2011). Resilience, ecology and adaptation in the experimental city. *Transactions of the Institute of British Geographers*, 36: 223–237.

Evans, J. and Karvonen, A. (2014). 'Give me a laboratory and I will lower your carbon footprint!'—Urban laboratories and the governance of low-carbon futures. *International Journal of Urban and Regional Research*, 38: 413–430.

Flood, C. (2013). Pension funds drawn to property. *Financial Times*, 13 January 2013.

Global Reporting Initiative [GRI]. (2015). *Sustainability Disclosure Database*. [Online]. Available: http://database.globalreporting.org [8 June 2014].

Gunder, M. and Hillier, J. (2009). *Planning in Ten Words or Less*. Farnham: Ashgate.

Hall, P.G. (2002). *Cities of Tomorrow: An Intellectual History of Urban Planning and Design in the Twentieth Century, 3rd Edition*. Oxford: Blackwell.

Hall, T. and Hubbard, P. (1996). The entrepreneurial city: new urban politics, new urban geographies? *Progress in Human Geography*, 20: 153–174.

Harvey, D. (1989). From managerialism to entrepreneurialism: the transformation in urban governance in late capitalism. *Geografiska Annaler B*, 71: 3–17.

Hodson, M. and Marvin, S. (2010). Urbanism in the anthropocene: ecological urbanism or premium ecological enclaves? *City*, 14: 298–313.

HSBC (2014). *HSBC Holdings plc – Environmental Efficiency*. [Online]. Available: http://www.hsbc.com/citizenship/sustainability/environmental-efficiency [27 February 2014].

Joss, S. (2010). Eco-cities – a global survey 2009. *WIT Transactions on Ecology and the Environment*, 129: 239–250.

Joss, S. and Molella, A.P. (2013). The eco-city as urban technology: perspectives on Caofeidian International Eco-City (China). *Journal of Urban Technology*, 20: 115–137.

Karvonen, A. and van Heur, B. (2014). Urban laboratories: experiments in reworking cities. *International Journal of Urban and Regional Research*, 38: 379–392.

Keeton, R. (2011). *Rising in the East – Contemporary New Towns in Asia*. Amsterdam: SUN Architecture.

Kjær, A.M. (2004). *Governance*. Malden, MA: Polity/Blackwell.

KPMG (2013). *The KPMG Survey of Corporate Responsibility Reporting 2013*. Amsterdam: KPMG International.

Olds, K. (2001). *Globalization and Urban Change: Capital, Culture, and Pacific Rim Mega-Projects*. Oxford: Oxford University Press.

Principles for Responsible Investment [PRI]. (2015). *About the PRI Initiative – Principles for Responsible Investment*. [Online]. Available: http://www.unpri.org/about-pri/about-pri [8 June 2015].

Rapoport, E. (2014). Utopian visions and real estate dreams: the eco-city past, present and future. *Geography Compass*, 8: 137–149.

Rapoport, E. (2015). Globalising sustainable urbanism: the role of international masterplanners. *Area*, 47: 110–115.

Saunders, T. (2008). *A Discussion Document Comparing International Environmental Assessment Methods for Buildings*. Glasgow: BRE Global Ltd.

Shaw, K. (2013). Docklands dreamings: illusions of sustainability in the Melbourne Docks Redevelopment. *Urban Studies*, 50: 2158–2177.

US Green Building Council [USGBC]. (2014). *Projects Directory*. [Online]. Available: http://www.usgbc.org/projects/neighborhood-development [9 June 2014].

Williams, K. (2009). Sustainable cities: research and practice challenges. *International Journal of Urban Sustainable Development*, 1: 128–132.

Wu, F. (2007). Re-orientation of the city plan: strategic planning and design competition in China. *Geoforum*, 38: 379–392.

7 Cabin ecologies

The technoscience of integrated urban infrastructure

Simon Marvin and Mike Hodson

Introduction

In October 2013, Cupertino City Council approved Apple's plan for a new campus. The new campus is on a 0.71 km^2 site and will house up to 13,000 employees in one central, four-storied circular building of approximately 260,000 m^2. The building will be surrounded by extensive landscaping, and underground parking. Media reports widely described the new structure as a 'spaceship' (Green 2014). Stefan Behling, one of the Foster and Partners architects working on the project, argues that the campus is 'one of the most environmentally sustainable projects on this scale anywhere in the world' (Vanhemert 2013). The spaceship analogue is not solely based on the aesthetics of the circularity and high-tech finish of the design but is also linked to the large-scale applications of technologies and techniques that were developed in the context of the military and space programmes. The campus will use advanced energy, water and waste treatment technologies to allow the development to be largely self-reliant and independent from centralised grids. The plan is for the facility to run entirely on renewable energy, drawing largely from on-site fuel cell plants and rooftop photovoltaic arrays.

The motivation for producing this chapter is to understand the history of the technoscience of the production, circulation and deployment of integrated urban infrastructure. The starting point for this project was the striking similarities between the forms of representation of metabolic flows of resources and integrated infrastructures in the design of spacecraft cabins and the design of green buildings. The chapter seeks to develop an understanding of how infrastructural integration appeared to be based on a set of rational and standardised techniques designed to abstract, audit, quantify, manage and reintegrate flows of resources – water, energy, wastes and carbon – at a number of different levels (individual/astronaut, building/space compartment and even the city) (Cosgrove 1994; Leslie 2006; Höhler 2008; de Mochaux, 2011). What struck us was how little understanding of the history of these techniques and approaches we had as urban infrastructure specialists and yet how potentially important they might be in framing the purpose and processes of integration. Here the work of historians Peder Anker (2005, 2010) on the development of ecological architecture and Jennifer Light (2002, 2003) on the urban interventions of defence intellectuals in the United States in the 1960s and 1970s is critical in providing a much more nuanced and sophisticated understanding of the technoscience that sits behinds contemporary urban interest in integrated and enclosed ecologies. This work reveals the tight connections between the production of 'cabin ecologies' – enclosed artificial environments that can sustain life in space through the development and application of integrated infrastructures – and the ecological design movement that is realising self-sustaining settlements using these same integrated techniques.

Using this insight into the connections between space and military programmes to construct 'cabin ecologies' for capsules, biomes, spaceships, space settlements and submarines, this chapter explores the production, dissemination, testing and deployment of integrated urban infrastructure technoscience, and the ways in which the boundaries between the interior and exterior environments are being selectively but significantly redrawn (Graham and Marvin 2001; de Cauter 2004; Sloterdijk 2009; Klauser 2010; Steinberg *et al.* 2011). Our concern is that an important shift has occurred in which the logic of the cabin ecology is emerging as a key response to fear about security and ecological change (de Cauter 2004). De Cauter (2004: 29) captures this when he talks about processes of encapsulation and the cellular city: 'capsule architecture is the architecture of the new city. It is architecture that functions like a space capsule. That creates an artificial ambient atmosphere, minimises communication with the outside and forms an isolated environment of its own.' As Peter Sloterdijk (2009: 142) writes in 'Foam City', the process of constructing enclosed habitats goes beyond the historical forms of city parks and greenhouses:

> The *Entkapselungmotiv* [process of encapsulation] comes to operate at such a scope that it integrates larger and previously external landscapes. The modern city (and urban landscape) develops more and more into an operational unit of the sprawling triad of space station, greenhouse and Human Island.

These enclosed artificial environments are themselves connected as an archipelago of enclosures through networks of capsules – the train, the automobile, the aeroplane and the space capsule (de Cauter 2004: 45). What we see here is the extension of the ecological dimension of splintered urbanism (Graham and Marvin 2001) where rebundled synthetic environments not only include integrated infrastructures but also other urban life support systems of atmospheric, temperature and climate control (Sloterdijk 2009). Increasingly, cabin ecologies become about protecting the inside from the threats and dangers of outside environments (Hodson and Marvin 2009).

This chapter is divided into five sections. First, we provide a brief overview of the wider technoscientific contexts within which cabin ecologies were developed, to illustrate the ambition, scale and scope of the international effort to build enclosed ecologies. Second, we identify the key features of the technoscience of integrated infrastructure that provides the basis for the technologies and practices of integration across infrastructure in engineering, design, architecture and urbanism. Third, we review the emergence of the new science of biospherics and the experimental spaces created to develop enclosed ecologies. Fourth, we examine the ways in which the concept of cabin ecologies is being applied to cities more widely to build enclosed life support systems on earth. In conclusion, we argue that urban infrastructure studies need to take the hidden history of infrastructure and enclosed life support systems more seriously in a period when anthropogenic change is reshaping the global ecological context of urban life, creating responses that seek to rebundle selected ecologies and infrastructure to ensure urban reproduction.

'Enclosed' life support systems

The origins of cabin ecologies can be found in the attempts to construct enclosed life support systems to deal with stressed ecologies (Colomina *et al.* 2004; Cohen 2011). Increasing attention on exploiting the subsea, deserts, the Arctic and the upper atmosphere, as well as national space exploration operating beyond the earth's atmosphere, led to the development of research and development programmes to create enclosed life support systems focused on three challenges.

Developing cabin ecologies

Military and space programmes built on already existing and partial knowledge of semi-enclosed cabins that provided temporary, short-term support for humans operating in military aviation and submarines (Cohen 2011). During the 1950s and 1960s, the desire was to operate in the external environments of space and the deep-sea, and this developed new challenges both in terms of the longer timescales required for operating in extreme environments as well as the scope and scale of support required to sustain human life. As the perceived operational need to occupy increasingly demanding and stressed environments of outer space and extreme environments was developed in cold war competition, the need for more demanding and sophisticated cabins was established. As programmatic understanding of the real constraints and serious danger of 'exterior' stressed environments to human life – which increasingly stretched beyond planetary limits – grew, the need also to develop new technologies and knowledge about 'interior' environments of the cabin was also being prioritised. The requirements for colonising and occupying – even if only temporarily – stressed environments drove an increasing focus on the possibilities and potentials of new types of inside environments. These cabin ecologies could more effectively and efficiently sustain human life in submarines, capsules and other forms of cabin ecology that had to operate under severe infrastructural and atmospheric constraints.

Enlarging the concept of infrastructure to include life support systems

The operational requirements to place humans in stressed environments for longer than a few hours increasingly pushed cabin ecologies programmes into developing new knowledge about human metabolic needs and how these could be sustained. Building interior ecologies increasingly required the concept of a human life support system that could provide not only the conventional modes of infrastructure of energy, water, mobility, and waste disposal but also had to look much more widely at sufficient heating and cooling, air quality and the generation of oxygen and disposal of carbon, and the need for food and water for drinking. The 'interior' cabin ecology was increasingly being asked to work as a total human life support system, with both the infrastructures of everyday life and the ecological services of clean air, heating and cooling needing to be reconstructed and replicated within the tight confines of an enclosed and partially autonomous cabin that was no longer physically connected to network infrastructures or earth-based ecologies. This cabin ecology had to recreate, at least partially, enclosed and quasi-autonomous or largely self-reliant infrastructures and ecologies of everyday life to sustain human life in enclosed cabins.

Developing new 'integrated' expertise, knowledge and technology

The critical challenge was then to create the expertise, knowledge and technology to realise enclosed spaces that could selectively rebundle infrastructures and life support systems for (temporary) human survival beyond centralised infrastructures and support systems. This aspiration touched on almost every aspect of social science, science and technology because the programmes needed to recreate environments and atmospheres that functioned much as Earth did. Consequently, the complex infrastructures of a typical city and the ecological support services had to be selected, miniaturised, organised and operated under the extremely demanding conditions of the cabin ecology. Programmes sought to integrate a wide range of expertise and technologies in this effort. Central to these were engineers, ecologists, psychologists, and town planners who were brought together in large interdisciplinary programmes to develop the technologies and techniques of cabin ecologies.

In summary, the primary task was threefold: to develop a new integrated understanding of different life support systems; to construct integrated responses using minimal resources by exploiting recycling and reuse strategies in the cramped and demanding operating conditions of space; and to produce systems that could operate securely and reliably off grid and in exposed environments. Overall, a specific set of knowledge and technology was developed that was to have important implications for integrated infrastructures.

Producing cabin ecologies

The technoscience of integrated infrastructure and life support involved the merging of scientific knowledge with industrial technologies in attempts to reshape resource flows for life support in new and more efficient ways. The developers of these programmes saw themselves not only as 'primarily engaged in subjugating nature and its processes through creating artificial natures via technological artifacts and systems, but through designing and engineering nature in the sense of reshaping and improving it' (Weber 2006: 403). The technoscientific project was distinctive because it combined different types of disciplinary expertise to integrate existing scientific knowledge in a manner that enabled the understanding and re-creation of life support systems in a miniaturised form. There were three key steps involved in the production of cabin ecologies.

Metabolic circulation

The first step involved the development of new forms of understanding about the metabolic needs of the human body. New knowledge was needed about the inputs of energy, food, water and atmospheric requirements to sustain life and the metabolic processes of the transformation and disposal of these resources. Drawing upon knowledge and expertise from across existing disciplines, there were efforts to develop an integrated understanding of the different resource requirements of the human body and the minimum requirements to sustain life (see Figure 7.1). Second, new techniques were needed to model these flows at different levels, from the individual body to groups of humans within enclosed ecologies of the cabin. Working with ecologists, new techniques were developed for modelling and representing resource flows through the use of input–output models. Drawing on systems diagrams from the electricity sector, ecologists used similar techniques to represent resource flows of water and energy. These techniques provided ways of mapping, measuring and manipulating resource

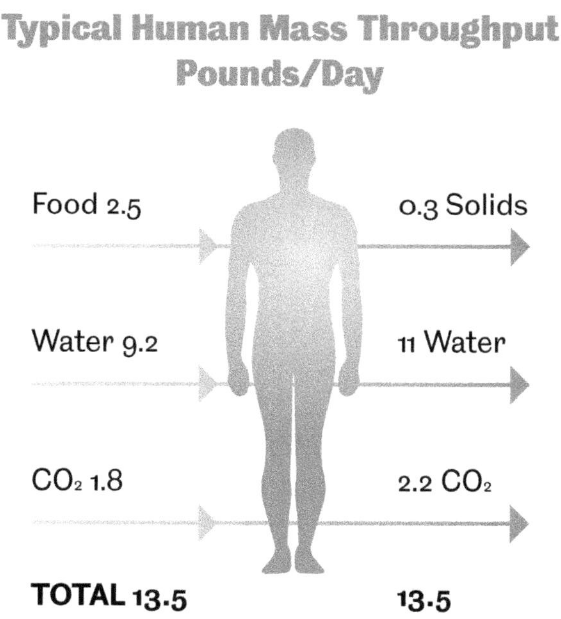

Figure 7.1 Human metabolic flow (source: NASA).

flows within cabin ecologies. Third, the developers of cabin ecologies were very interested in the concept of 'carrying capacity', which had been previously used to measure the maximum number of people allowed on a ship or ferry. This was re-developed into a biological concept that identified the maximum population size that an artificial environment of food, water and energy could sustain over a particular time period (Sayre 2008). The search was for a stable long-term solution to providing life support in cabin ecologies. Through these techniques, a 'metabolic' approach to cabin ecologies was developed that would later strongly influence thinking about potential applications of integrated infrastructures in urban contexts.

New technologies

Alongside the focus on metabolic flows, there were large programmes of investment in new technologies for supporting cabin ecologies. These programmes focused on the development of fuel cells and photovoltaic electricity technologies, water purification and production, recycling and reuse of wastes, air purification and removal, and storage of different wastes. These environmental technologies are now at the core of integrative infrastructural technologies (Zimmerer 1994).

The human body became a focus for technological and biological enhancement for the stressed conditions of space. NASA sponsored much of the early work on the concept of the cyborg body just 3 years after the term was developed in 1960 (Driscoll 1963). Humans were biologically designed to operate within the parameters of the Earth's environment and there were many areas where human capabilities fell short of 'mission requirements'. The NASA report (Driscoll 1963) outlined a number of areas where human functionality could be

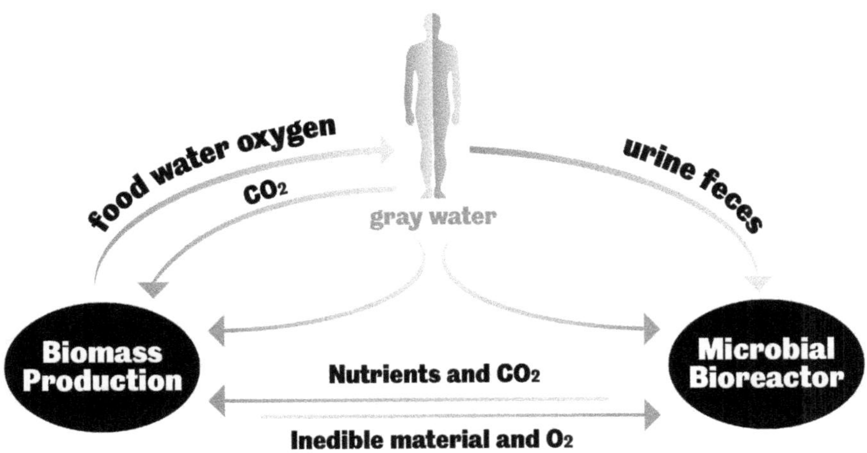

Figure 7.2 Integrating humans into life support systems (source: NASA).

enhanced through cyborg efficiency (artificial organs, drugs, hypothermia and sensory deprivation, for example). Although the concept of the cyborg was not taken forward in any significant way, the human body was recognised as having a key role within the integrated systems both as a source of inputs but also producing wastes – liquids, gases and solids – that could potentially be incorporated into life support systems. Figure 7.2 illustrates the ways in which human metabolic needs and wastes started to become integrated with food, water and oxygen systems.

Cybernetic systems

Cabin ecologies themselves were increasingly reframed to emphasise the management of processes. Adopting a language of feedback, homeostasis and control, life support systems were seen as cybernetic systems requiring systems analysis and computer simulations as problem-solving tools (Edwards 1996). The cabin became redefined in cybernetic terms by unifying different traditions that understood the cabin as both organic systems and as machines. Cybernetic systems offered an opportunity to merge understandings of the cabin from the biological and physical sciences with those of mechanical systems and organisations. Systems analysis and computing could knit these different visions of the city together (Light 2003). Key to this was the incorporation of action and feedback in tools such as databases and computer simulations where real-time information could properly represent the circulation and flows of resources in ways in which models could not. Figure 7.3 illustrates the ways in which humans and organic systems as well as machines and control systems were integrated into a total life support system.

The technoscience of cabin ecologies was critical in understanding the metabolic system of the body and enclosed ecologies by developing technologies for providing infrastructure and re-assembling these into new integrated infrastructural configurations.

The new discipline of biospherics

Huge technoscientific effort was put into constructing cabin ecologies through development and demonstration. These new assemblages attempted to reconfigure essential life support through a series of selected windows that illustrate tensions, controversies and contradictions about how life could be sustained through essential infrastructure. A new scientific object of study was to test and develop cabin ecologies: materially closed ecological systems (CES) supported by a new discipline – biospherics. This is an 'integrative' discipline – drawing on biology, physiology, ecology, microbiology, engineering and social sciences – that sought to integrate technology, nature and humans:

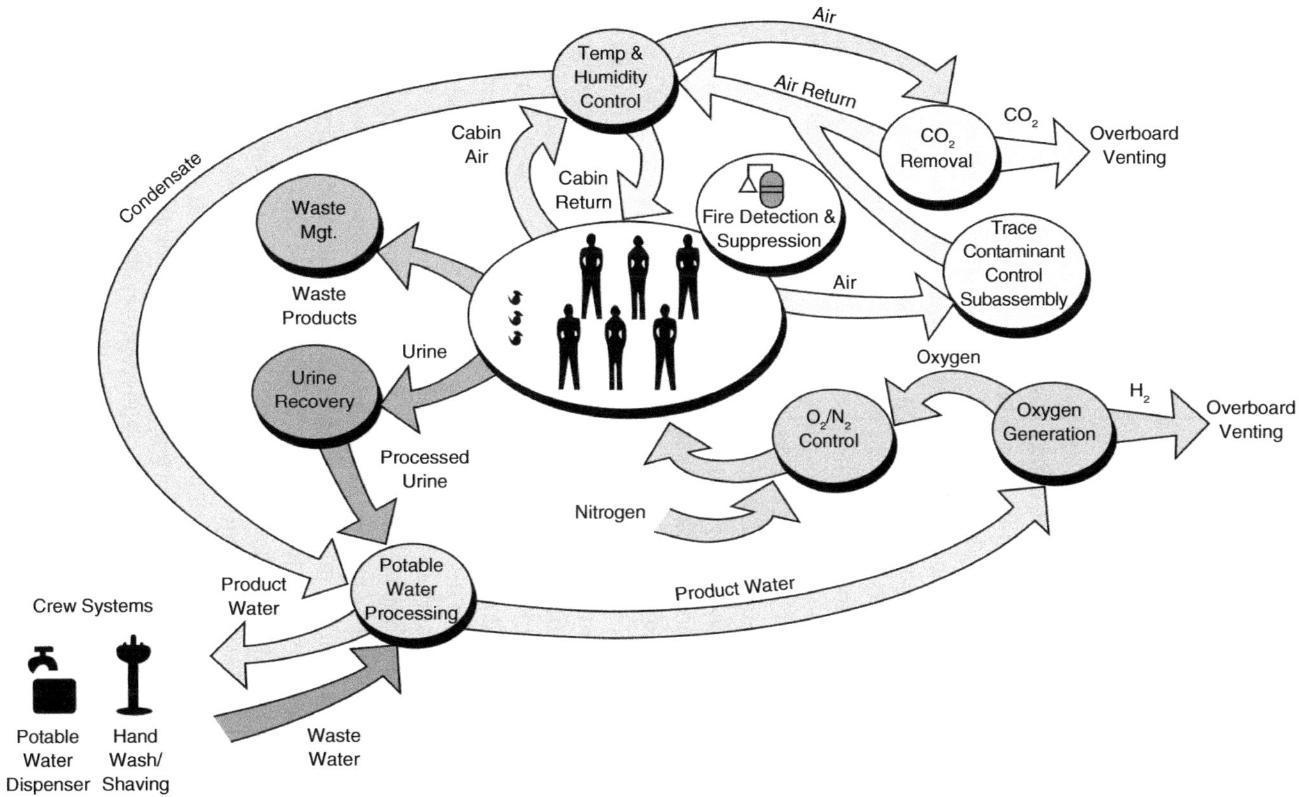

Figure 7.3 International Space Station as a cybernetic system (source: NASA).

> To design, construct, and study artificial 'biospheres,' it is necessary to intelligently design and manage the biogeochemical flows of matter and energy, to use sophisticated technologies and computer/information systems, to incorporate the achievements of genetics, biotechnology, and bioengineering and to make use of time-tested and reliable natural ecological mechanisms.
>
> (Nelson *et al.* 2010: 518)

Different types of artificial environments were created that initially focused on ecospheres, small lab-based systems that are materially closed but open to energetic input from sunlight or artificial light. These are not designed to support human life but permit the laboratory study of small ecosystems. However, biospherics was focused on the development of bio-regenerative technologies – biological systems enhanced by other technologies – to provide life support services of food, air and water. An example of a bio-regenerative technology is a plant chamber in which a crop regenerates part of its atmosphere, purifies water through transpiration and produces food. These technologies are key components of controlled or closed life support systems. 'Controlled' life support systems utilise stored resources of food, air and water and physiochemical means to handle wastes, as these are only partially regenerative systems. 'Closed' life support systems are ecological systems that are materially closed – recycling all their major elements – but energetically (and informationally) open or they would decline due to increasing entropy. The light needed from photosynthesis is supplied by artificial light powered via solar arrays or nuclear energy.

There are currently nine experimental ecosystems used to conduct research with closed material cycling. These are highly abstract, technological environments that attempt to rebundle humans, technologies, gases, materials and ICT to create life support systems for humans. As Luke (1997: 108–109) argues, enclosed ecologies included 'only those life-forms deemed essential to the reproduction of human life in an artificially encapsulated sphere of material and energy flow'. Selected fragments of nature are held together inside an 'ecological formation in which humans, computers, mechanisms and biomasses become one interdependent coevolutionary energy generation and conversion circuit'. For example, three of the most important objectives of the new NASA Advanced Life Support Project (2015) are: (1) to develop life support systems that 'significantly reduce life cycle costs, improve operational performance, promote self-sufficiency, and minimise expenditure of resources for long-duration missions'; (2) to 'develop and apply methods of systems analysis and engineering to guide investments in technology, resolve and integrate competing needs, and guide evolution of technologies'; and (3) to 'transfer technologies to industrial and residential sectors for national benefit'.

Figure 7.4 Biosphere 2 (source: Biosphere 2, University of Arizona).

Yet even the experience of the most complex engineering environments has been problematic. Biosphere 2 (Figure 7.4) was a private initiative designed to exploit commercially the technologies of enclosed ecologies but it collapsed when the internal environment could no longer support human life. Höhler (2008: 78) graphically shows how 'a drastic decline in pollinating insects and losses of other species had to be taken into account, and the slow decline of oxygen called for an oxygen injection one and a half years into the experiment' – the only environment in which it was possible to survive was outside and not in an artificially created synthetic interior. Despite this experience, 'what is being proposed for conquering space, is in fact, echoing the means being used now for recolonising any place on Earth' (Luke 1997: 111).

Capsular urbanism

The technoscience of cabin ecologies created the conceptual basis for a new approach to the design of settlements – an environmental ethic based on the notion of the 'scientifically manageable astronaut' was established. This was exemplified by 'a series of technologies for managing waste, air, food, and energy, the space cabin thus came to represent the rationale and scientific way of ecological living' (Anker 2005: 539–540). The techniques and technologies of the cabin ecology had important implications for urbanists and designers seeking to develop more ecological sensitive cities and buildings that could live within

Earth's limits. The ways in which cabin ecologies shaped urban thinking about integrated infrastructure and human life support systems are reviewed through four windows.

Ecological urbanism

Peder Anker brilliantly captures the history of ecological architecture in the 1960s and 1970s and the ways in which it was heavily influenced by the ethic of the spaceship as well as the specific techniques of the technoscience of cabin ecologies (see also Höhler 2008; Kallipoliti 2008). Life in space came to 'represent the peaceful, rational, and environmentally friendly alternative to the destructive, irrational, ecological crisis down on earth' (Anker 2005: 527). The challenge for the majority of ecological designers was a question of 'adopting space technologies, analytical tools and ways of living'. During the 1960s, 1970s and 1980s, there was a series of experiments in building more closed autonomous ecological systems that struggled towards 'encapsulating buildings so that the inhabitants would be sheltered against the coming doom' (Anker 2005: 538).

Anker reviewed the work of leading ecological designers, including Buckminster Fuller, John Todd, the New Alchemists, Alexander Pike, Brenda and Robert Vale, Ken Yeang and many others, and situated their design projects within the context of the ecological techniques and technologies of the cabin ecology. Key to many of these designs, and the later 'eco cities' movement, was the study of buildings as a closed ecological system in analogy to the closed cabin ecology of the spaceship. The technologies developed for waste, air, food and energy management in the space cabin came to represent the best way of developing a rational and scientific way of ecological living. For example, Alexander Pike at Cambridge University formed a research group to investigate the relevance of cabin ecologies to architecture with the aim of developing autonomous buildings. Brenda and Robert Vale were part of this group and developed their concept of autonomous buildings based on these techniques. This architecture was designed to integrate waste, water, and air and heat technologies into a technological whole, using ambient solar and wind energy as well as human and household waste material.

(Urban) control rooms

The landscape designer Ian McHarg used cabin ecology research tools in a series of proposals to improve global landscape management in the late 1960s:

> McHarg fashioned the landscape designer as a cabin ecological engineer managing and surveying the environment in analogy to how NASA's Ground Control in Houston kept a close eye on the cabin ecological circulation of energy and materials within a spaceship.
>
> (Anker 2005: 535)

This turned out to be a prescient statement when the Rio de Janeiro Control Room (COR) opened in 2010 to monitor the weather, transport and wider circulations in the city. The interior of COR showed remarkable similarities with a NASA control room – one designed to control ecology of the space capsule and the other the ecology of the city. These similarities are not accidental; the design, the technologies and the operation of the COR were closely modelled on Ground Control in Houston. NASA insisted that the different private contractors active in the control room wear the same uniforms to engender a shared culture – a practice that has been replicated in Rio where the 35 public and private institutions represented in the control room share a common uniform. COR is even linked to NASA weather prediction systems as it seeks to minimise and prepare for the threat of landslides and infrastructural disruption following heavy rainfall. This is no longer an analogue – the logic of the scientifically and cybernetically managed cabin ecology has now been scaled to the entire city through COR.

Transfer of cabin ecologies (to Earth)

NASA has a strategic priority to transfer technologies from the space programme to application on Earth in which cities and buildings are particular important. Sustainability Base is a NASA building designed and constructed to LEED (US Green Building Council's Leadership in Energy and Environmental Design) platinum standards. The building uses NASA technologies 'originally engineered for space travel and exploration' and, as a working building, it provides an 'exemplar' for the future of buildings (NASA Sustainability Base 2015). A journalist for the magazine *GreenSource* writes:

> Between rooftop solar arrays and a solid-oxide fuel cell, the building supplies more energy than it needs and sends the surplus back into the Ames grid. From an efficiency perspective, space travel turns out to be a terrific idea lab for building performance on Earth. Technologies that ensure astronauts' survival at the International Space Station – from urine-recycling water filtration to computer controls that protect the indoor environment and anticipate energy needs – also happen to excel at less glamorous, workaday tasks like managing daylight and regulating energy use.
>
> (Anderson 2013)

NASA also has various programmes designed to promote the use of space technologies on Earth in buildings and cities. For example, NASA @ Home and City (2015) showcases the science and technologies developed within NASA programmes and their application on Earth. More significantly, NASA is sponsoring work on a new architectural curriculum for off-planet designed buildings that could provide the basis for a new unified architectural curriculum for buildings irrespective of their location on or off planet.

Urban biospheric envelopes

An urban biospheric envelope is a structure that encloses a large urban area under a single roof. The envelope is typically airtight and pressurised, creating a habitat that can be controlled for air temperature, composition and quality, typically due to an external atmosphere that is inimical to habitation. The skin of the envelope may capture water for later storage and treatment, and in a number of cases may be covered with photovoltaic film so that the skin generates electricity. While there are plenty of examples of purified urban environments that seek to protect themselves from dangerous exteriors, these tend to be at the level of the individual building, such as a private home or a gated community, a transport hub, or a sport stadium. Here, the focus shifts to forms of enclosure that seek to encompass the whole of a city, the central business district or a zone of a city – comprising a patchwork of existing enclosures and the spaces between them.

Architects, developers and engineering companies have developed a number of detailed proposals for these eco-envelopes, including: Buckminster Fuller's dome over New York; detailed proposals for the city of Winooski in Vermont in 1979; a proposal for a dome over the former diamond mine of Mir in Siberia; and in 2009 a proposal to cover the entire central business district of Houston, Texas with a regenerative envelope to protect the city from hurricanes and other impacts expected from a changed climate, as well as to produce electricity. More recently, there have been proposals for eco-envelopes in the city of Beijing, in response to the appalling air quality that is a significant threat to human health there. The UK architects Orproject (2015) proposed:

> The construction of an enclosed park within the city. The park houses a botanical garden, the air inside the park is clean, and temperature and humidity are controlled throughout the year. The buildings surrounding the park, which are connected to the controlled air system, can house apartments, offices and retail, but may also offer sports or medical facilities, which make specific use of the healthy air. The heating and cooling of the air is done through a ground source heat exchange system. Electricity for the project can be generated by solar cells integrated into the canopy surface.

The designers of the project explicitly link the concept to the analogue of the botanical garden – but in this new concept it is not just the ecology that is being protected but also the humans! The designer Sodhi sees the envelope as 'just an infrastructure project like building metro stations and parks – it's applicable in every dense, polluted metropolis … "bubbles" is about surviving climate change'. While debate about the ethics of abandoning the atmospheric commons went on in the architectural community, an international private school in Beijing actually implemented its own eco-envelope. Parents were extremely concerned about the poor air quality not only in the city but also when at school, leading to calls for air quality monitoring within classrooms, the gymnasium and even the school bus. Because of poor air quality, children might be kept inside for up to 15 days in a row and were unable to play or undertake sports activities outside. In response to parents' concerns, the school purchased and installed a US$5 million envelope and air filtration system to cover the outdoor play and sports spaces and spends US$150,000 annually on air filtration. Of course, these protected atmospheres are the premium ecologies and do nothing to improve air quality more widely in the city.

In summary, we can see how the concept of 'cabin ecologies' migrates from one context to the other: from military research and the experiments of NASA's space programme to ecological architecture, to urban control rooms, to demonstrator buildings of space technologies, and finally to urban biospheric envelopes that evoke off-planet space stations on Earth. What remains constant is a new environmental consensus in the form of a capsular ecology where new metabolisms and life support systems are integrated into cities and buildings through the replication of natural ecosystems in 'synthetic environments' that protect the inside from the outside.

Conclusions

There are three key implications of this chapter for understanding contemporary interest in experimental cities. First, there would be value in a more sustained engagement with the historians of science, technology and architecture in order to reveal more about the particular temporal dynamics and social and institutional context in which knowledge and technologies of integrated infrastructure and life support systems have been developed. Of particular interest here are the social assumptions that underpin much of this research in creating selective and synthesised ecologies to protect humans from stressed ecologies, and the extent to which these securitised assumptions have shaped the path dependency of integration techniques and technologies in multiple urban contexts. It would be helpful to understand more about the ways in which different options – complex ecologies versus simple ecologies in closed environments and ecological versus technological forms of treatment – were chosen and what consequences these have had for our contemporary understanding of integrated infrastructures.

Second, cabin ecologies were not just focused on infrastructure networks – energy, water and waste – but were also concerned with a wider set of life support systems including the atmosphere, heat/cold, carbon and oxygen, food, plants and different ecological biospheres. As cabin ecologies sought to replicate the infrastructural and life support systems of earth within an enclosed space, new knowledge was developed about the creation of 'total' environments for human life support. Under conditions of global anthropogenic change and the much more uncertain context for urban ecological reproduction this creates, it is likely that urban infrastructural studies will need to understand a more expansive notion of infrastructure as issues around temperature, food, and air quality are likely to be more fundamental to questions of urban reproduction. This might involve the development of new collaborations with urban ecologists, Earth scientists, and technologists – many of the experts involved in the new disciplines of biospherics – to ask critical questions about the limits, potentials and wider politics of capsular urbanism.

Finally, there is value in undertaking further work on new urban enveloping as emerging forms of atmospheric enclosure – the effort to re-construct a safe 'outside' atmosphere 'inside'. The construction of urban eco-domes is designed to ensure human reproduction in the context of the changed ecological conditions of the anthropocene. These aspire to provide protection from pollutants, heat/cold and wet/dry, and to ensure continuity in the more testing conditions of global ecological change. There are three areas for further research. The first area is the detailed in-depth and comparative analysis of urban domes and especially the politics of their implementation and the redrawing of boundaries between humans and cities. Is this a new politics of enclosure in a period of anthropogenic change? Second, how are these processes of boundary formation contested – through legal challenge, regulatory systems, through concerns about who decides what is inside and outside, what is left outside (damaged, dangerous, ecologically denuded or neutral spaces) – and how effective is enclosure. Third, potential alternatives to fortressing and enclosure – are there more relational and alternative ways of thinking of boundaries in the anthropogenic era based on other ways of seeing?

References

Anderson, A. (2013). NASA Sustainability Base. *GreenSource: The Magazine of Sustainable Design*, January 2013. [Electronic]. Available: http://greensource.construction.com/green_building_projects/2013/1303-nasa-sustainability-base.asp [5 June 2015].

Anker, S. (2005). The close world of ecological architecture. *The Journal of Architecture*, 10: 527–552.

Anker, P. (2010). *From Bauhaus to Ecohouse: A History of Ecological Design*. Baton Rouge: Louisiana State University Press.

Cohen, J.L. (2011). *Architecture in Uniform: Designing and Building for World War II*. New Haven, CT: Yale University Press.

Colomina, B., Brennan, A. and Kim, J. (eds) (2004). *Cold War Hot Houses: Inventing Postwar Culture from Cockpit to Playboy*. New York: Princeton Architectural Press.

Cosgrove, D. (1994). Contested global visions: one-world, whole-Earth, and the Apollo space photographs. *Annals of the Association of American Geographers*, 84: 270–294.

de Cauter, L. (2004). *The Capsular Civilisation: On the City in the Age of Fear*. Rotterdam: NAi Publishers.

de Mochaux, N. (2011). *Spacesuit Fashioning Apollo*. Cambridge, MA: MIT Press.

Driscoll, R.W. (1963). *Engineering Man for Space*. May 15 1963. The NASA cyborg study (NASw-512). [Online]. Available: http://cyberneticsoo.com/wp-content/uploads/2012/01/cyborg-nasa-driscoll-1963.pdf [5 June 2015].

Edwards, P.N. (1996). *The Close World – Computers and the Politics of Discourse in Cold War America*. Cambridge, MA: MIT Press.

Graham, S. and Marvin, S. (2001). *Splintering Urbanism: Networked Infrastructures, Technological Mobilities and the Urban Condition*. London: Routledge.

Green. J. (2014). Apple's spaceship campus: Steve Jobs' environmental legacy? *Greener Ideal*, 6 February 2014. [Electronic]. Available: http://www.greenerideal.com/alternative-energy/0206-apples-spaceship-campus-steve-jobs-environmental-legacy/ [5 June 2015].

Hodson, M. and Marvin, S. (2009). Urban ecological security: a new urban paradigm? *International Journal of Urban and Regional Research*, 33: 193–215.

Höhler, S. (2008) 'Spaceship Earth': envisioning human habitats in the environmental age. *GHI Bulletin*, 42: 65–85.

Kallipoliti, L. (2008). Feedback man. *Log*. 13/14: 115–118.

Klauser, F.R. (2010). Splintering spheres of security: Peter Sloterdijk and the contemporary fortress city. *Environment and Planning D: Society and Space*, 28: 326–340.

Leslie, T. (2006). Just what is it that makes capsule homes so different, so appealing? Domesticity and the technological sublime, 1945 to 1975. *Space and Culture*, 9: 180–194.

Light, J. (2002). Urban security from warfare to welfare. *International Journal of Urban and Regional Research*, 26: 607–613.

Light, J. (2003). *From Warfare to Welfare: Defense Intellectuals and Urban Problems in Cold War America*. Baltimore, MD: John Hopkins University Press.

Luke, T.W. (1997). *Ecocritique – Contesting the Politics of Nature, Economy, and Culture*. Minneapolis: University of Minnesota Press.

NASA Advanced Life Support Project (2015). NASA Advanced Life Support website. [Online]. Available: http://www.nasa.gov/centers/ames/research/technology-onepagers/advanced-life-support.html [5 June 2015].

NASA @ Home and City. (2015). NASA @ Home and City website. [Online]. Available: https://www.nasa.gov/externalflash/nasacity/index2.htm [5 June 2015].

NASA Sustainability Base. (2015). NASA Sustainability Base website. [Online]. Available: http://www.nasa.gov/ames/facilities/sustainabilitybase [5 June 2015].

Nelson, M., Pechurkin, N.S., Allen, J.P., Somova, L.A. and Gitelson, J.I. (2010). Closed ecological systems, space life support and biospherics. In Wang. L.K., Ivanov, V., Tay, J.-H., and Hung, Y.-T. (eds), *Environmental Biotechnology*. New York: Humana Press, 517–565.

Orproject. (2015). Orproject website. [Online]. Available: http://orproject.com/bubbles/ [5 June 2015].

Sayre, N. F. (2008). The genesis, history, and limits of carrying capacity. *Annals of the Association of American Geographers*, 98: 120–134.

Sloterdijk, P. (2009). Foam city: about urban spatial multitudes. In Ramos, S., Allen, S., Boeri, S. and Turan, N. (eds), *New Geographies*. Cambridge, MA: Harvard University Press, 136–143.

Steinberg, P.E., Nyman, E. and Caraccioli, M.J. (2011). Atlas swam: freedom, capital, and floating sovereignties in the seasteading vision. *Antipode*, 44: 1532-1550.

Vanhemert, K. (2013). Look inside Apple's spaceship headquarters with 24 all-new renderings, *Wired*, 11 November 2013. [Electronic]. Available: http://www.wired.com/2013/11/a-glimpse-into-apples-crasy-new-spaceship-headquarters [5 June 2015].

Weber, J. (2006). From science and technology to feminist technoscience. In Davis, K., Evans, M. and Lorber, J. (eds), *Handbook of Gender and Women's Studies*. London: Sage, 397–414.

Zimmerer, K.S (1994). Human geography and the 'New Ecology': the prospect and promise of integration. *Annals of the Association of American Geography*, 84: 108–125.

Part II

Experimenting in cities

8 Green enclaves, neoliberalism and the constitution of the experimental city in Santiago de Chile

Martin Sanzana Calvet and Vanesa Castán Broto

Introduction

To what extent are experiments in urban sustainability a means to re-imagine and adapt neoliberal urbanism? In Santiago de Chile, experiments based on innovation in urban policies, governance, planning and urbanism have led to the development of residential enclaves characterised by a discourse of greenness and sustainable urban development. These experiments in urban sustainability have emerged after the political transition from dictatorship and the stabilisation of a democratic regime in Chile during the 1990s, a period during which neoliberal policies were contested, adapted and consolidated. Neoliberalism is a widely used and contested term in current urban theory, but also one that is open to wide interpretive flexibility. In relation to the city, its lived experiences and how it is governed, neoliberalism is often represented as an ideology whereby the privatisation of public resources has become axiomatic (Hackworth 2007). However, neoliberalism has to be considered in a new light as a combination of two parallel processes: one in which the state is made responsible for the preservation of the market and competition rather than for maintaining the well-being of its citizens; and another whereby citizens themselves adjust social relations in such a manner that they can be integrated in those markets (Schipper 2014).

Positive readings of experiments in urban governance have argued that they can be directed towards the reimagination of spaces of possibility for more progressive and more sustainable societies, building upon the spaces of hope that emerge from contingent and place-based analysis of social interactions (Coutard and Guy 2007). Yet, if we take the neoliberal project as a key discourse in the shaping of contemporary cities, the experimentation project needs to be examined in relation to specific contexts of neoliberal urbanisation. What happens when experimental, future-oriented perspectives encounter the real-politik desert of neoliberal urbanism? To date, analyses of such an encounter have focused on the extent to which green-inspired progressive ideals are, in practice, captured by neoliberal ideologies. As argued by While and colleagues (2010) and Jonas and colleagues (2011), state projects of carbon control, for example, are often found within broader neoliberal programmes. Moreover, there are no guarantees that the impacts of well-intentioned environmental interventions can do anything more than simply reproduce neoliberal strategies for urban and regional management. In this case, experimentation may be a means for the reproduction of the neoliberal city rather than a mechanism whereby new forms of engagement with the built environment and the social structure in cities can be introduced. Experiments are a site for potentiality, and yet, this very potentiality raises questions about the kinds of change they can foster. There are no guarantees that changes will follow experiments and even less that these changes will be inherently progressive. Thus, we can ask if neoliberal spaces are particularly fertile for experimentation and if these experiments serve as mechanisms for the reproduction of the neoliberal order. The proliferation of sustainability experiments in enclaves in the peri-urban area of Santiago de Chile is an opportunity to examine these questions, as they emerge within a particular urban context where neoliberal doctrine has been enshrined but perhaps never fully realised since the dictator Augusto Pinochet gained control of the country in 1973.

Experimentation and the production of urban environments

The notion of experimentation in global environmental governance has emerged in association with socio-technical analyses of urban environments. These ideas have been particularly prominent in debates about cities and climate change, and characterise an identifiable reaction against a generalised discontent with the lack of progress in the international climate change regime which has encouraged a diverse range of public, private and civil society actors to develop a variety of transformative initiatives (Bulkeley *et al.* 2014). These ideas can be situated within the framework of a broader understanding of experimentation concerning processes of spatial governance and the need to re-imagine socio-ecological relations in the city, linked to a thriving literature on urban laboratories (Karvonen and van Heur 2014). A central theme in this literature is to separate the analysis of urban governance from what is perceived as a dominant current of neo-Marxist urban thought that emphasises the structural drivers of socio-spatial inequality and constitute urban space as inherently exploitative (Karvonen and van Heur 2014). So while STS scholars during the 1990s called for the need to bring scholarship 'outside the laboratory' to examine the workings of science in quotidian spheres of power (e.g. Gieryn 1999), contemporary analyses of urban laboratories bring the city inside the laboratory, and use the laboratory as a lens through which to study the city itself.

In relation to normative ideas about the roadmap towards a more sustainable society, experimentation is regarded as a key mechanism whereby socio-ecological transformations can be attained. Much of this debate has framed transitions to sustainability as processes that entail a radical reconfiguration of dominant regimes; that is, the assemblages of materials, meanings, policies and tacit rules of conduct that shape both innovators' and users' relations to technologies and resources and that make socio-material relations particularly obdurate (Markard *et al.* 2012). In relation to cities, a key concern has been how to diagnose such change and how to bring it about in a context in which multiple socio-technical systems may simultaneously interact and where conceptual heuristics do not match the multiplicity of relations that characterises urban experiences and imaginaries (Hodson and Marvin 2010). Another key concern is the extent to which experiments are truly able to create action spaces and operate independently from the broader regimes in which they emerge (Bulkeley *et al*. 2014). Ultimately, experiments depend on the grammar of regimes within which they are constituted.

Let us imagine that there are indeed, as Karvonen and van Heur (2014) argue, two schools of thought on urban change: one concerned with the structural drivers of socio-spatial inequalities which would have David Harvey as its main figurehead; and one related to experimentation and the place-based reimagination of spaces of possibility, which follows the thought of Actor–Network Theory and has Bruno Latour as its main referent. It is curious to observe how both schools, as separated as they may be, have actually come together in asking a very similar question. So while neo-Marxist scholars have long been asking the question about whether progressive action is actually possible outside the dominant outfit of capitalism (dressed up as neoliberalism in contemporary times), STS scholars are concerned with the extent to which true innovation can emerge from within the dominant configurations of regimes. In practice, both perspectives resonate with urban laboratories, where spaces of possibility are created at the same time as they are entrenched in narratives of growth and competition (Evans and Karvonen 2014). Here, there is an opportunity to consider the extent to which the contradictory characteristics of experimentation enable its constitution as a form of future governance (Castán Broto 2015).

One approach that has attempted to bridge the perceived divide between these schools of thought has been urban political ecology (UPE). UPE analyses reflect upon the city as a site of socio-ecological transformation, where different material and symbolic exchanges relate to the political materialisation of hegemonic understandings of what the city is or ought to be. More broadly, this literature explains technological development in urban contexts as embedded in a process of nature domination in which technological innovation follows the need to adjust power structures to the evolving demands of the urban environment (Swyngedouw 1997, 2004). From this perspective, and in relation to the dependence of sustainability experiments from dominant regimes of capitalist reproduction and neoliberal discourse, UPE provides two fundamental insights for the analysis of experiments. The first one is a realisation that experimentation is a fundamentally ambiguous process that is not inherently progressive, but, equally, whose results cannot be simply subsumed within a pessimistic sense of assimilation by dominant regimes. The second one is that experimentation needs to be situated within the specific urban context in which experiments emerge, and in relation to the actual material work that they perform within a given urban context.

Bulkeley *et al*. (2014) have developed a framework for the analysis of experiments that uses this perspective both to situate experiments within the context of governance in which they are produced and to assess them in relation to their actual impact on that urban fabric. They propose that experiments can be analysed, first, in relation to how they are made. This involves the assembly and arrangement of a series of discursive and material elements to make the experiment possible as a means to structure the experiment and explain how it should be done but also make it compelling and, hence, necessary. Second, experiments should be explained in relation to how they are maintained; that is, how they need to be adjusted to a particular urban context but, also, the extent to which they are able to readjust urban relations to maintain stability while reproducing the discourse of innovation that makes them necessary. Third, experiments should be analysed in relation to how they are lived; that is, how they contribute and are appropriated within specific urban practices so that they become an essential part of life in the city.

Our methodology involves reflection upon what actually constitutes an experiment beyond specific initiatives and technologies. Following Bulkeley *et al.* (2014), the experiment here is characterised by the relation between the intention to provoke a certain change in a normative direction and the lack of understanding of how precisely the intervention will provoke such a change. In this sense, experimentation requires the development of a rationale for such experimentation as well as an explanation of how the experiment fits in its spatial and governance context. This is what Bulkeley and colleagues (2014: 228) describe as 'making the experiment calculable and compelling'. The processes of realising the relationships between intervention and results constitute experimentation.

The following sections present an analysis showing how the proliferation of state- and market-led sustainability experiments in enclaves in the peri-urban area of Santiago de Chile, aimed to increase urban sustainability and the greenness of urban development, evolved within a contested framework of neoliberal policies. Experiments are intended to address global terrors as well as cope with everyday concerns about environmental resources and pollution threats (Hodson and Marvin 2010). While experiments may not be a project of neoliberalism, they emerge within a neoliberal system and may help to oil the system for its reproduction. In Santiago de Chile, enclaves are the preserve of a wealthy class that seeks to escape urban hassles by inhabiting privileged spaces (cf. Davis 1998; Caldeira 2000). Yet, these residential enclaves are clearly linked to a specific discourse of greenness and natural living at the edges of the city. There is thus an urgent need to understand the extent to which the modality of eco-experiments in Santiago – the enclaves – helps to reproduce neoliberalism by adjusting to changing conditions while simultaneously creating new and compelling discourses of greenness.

Methodology

To summarise our case, the empirical analysis provides an overview of the ideological and regulatory context of enclaves and a systematic analysis of how enclaves are made, maintained and lived in. This socio-technical assemblage of planning innovation, green discourses and technological interventions constitutes an experimental attempt to reconfigure the socio-ecologies in which neoliberal urbanisation is being realised in Santiago. In particular, our analysis of the development of green enclaves focuses on a group of residential enclaves developed by a singular planning innovation enacted in 1997 for a specific rural area of Santiago metropolitan region, called Chicureo, where the government attempted to contain urban sprawl by boosting the emergence of new self-sufficient and sustainable urban centres which are fully funded, designed and operated by private developers. The analysis emphasises how these experiments combine state-led urban entrepreneurialism with housing market-led sustainability.

To capture the complexity of this new and greener form of urbanisation, we used multiple methods that describe the scales and perspectives of the enclave phenomenon. Data for the analysis were derived from 56 semi-structured interviews with relevant actors in the production and reproduction of the enclave, from the state, the private sector and civil society;[1] an online survey conducted with 47 enclave residents;[2] a review of the marketing materials in a selected housing market magazine;[3] a review of the official documents of the metropolitan master plan enacted in 1994 and its modifications until 2014; a review of the resolutions of the official environmental impacts assessments for different enclaves; and photographs and notes taken in the field.[4] The analysis reveals the role of planning innovation in making the experiment, the insertion of the enclaves through experiment maintenance, and how living in the enclaves has consolidated certain processes of subjectification.

The making of green enclaves as neoliberal planning innovation

To tackle Santiago's unequal growth and distance themselves from the radical neoliberal policies of the 1980s, the first post-dictatorial authorities conceived a new urban master plan for metropolitan Santiago (PRMS) to regulate the urban land use through zoning, densification and mixed land use, while limiting sprawl by surrounding the city boundaries with a green belt of agricultural and protected natural areas (Petermann 2006). This master plan, enacted in 1994, triggered unexpected outcomes, such as a speculative boom of rural land subdivisions in the countryside adjacent to Santiago by landowners who feared further restrictions and by affluent land buyers aiming to secure a plot of rural land – known as parcela de agrado[5] – for residential and leisure purposes (Hidalgo *et al.* 2005).

Critiques of the PRMS and its supposed failure to reduce sprawl and the loss of agricultural land prompted a coalition of technocratic authorities and the property development industry to campaign for renewed entrepreneurialism in urban policies. The Ministry of Housing pushed for a modification in the PRMS to allow for new modalities of neoliberal urbanism (Poduje and Galetovic 2006). Instead of preventing sprawl towards rural areas, the modification to the PRMS enacted in 1997 attempted to govern it, by annexing the whole rural province of Chacabuco to the master plan, subjecting it to an ad hoc planning innovation based on regulated neoliberalism. The rationale of the 'Plan Chacabuco', as it was known colloquially, was to protect the rural and wild environment of the area and increase the sustainability of Santiago's growth by promoting the development of new 'self-sufficient cities' fully developed by the private sector that would contribute to a polycentric, sustainable and less segregated development of Santiago (MINVU 1997). The main innovation of the plan was to establish a limited number of sites for urban expansion in the Province of

Chacabuco, called Zones of Conditioned Urban Development (ZUDC). These large rural estates of at least 300 hectares were re-zoned for urban residential development with higher housing densities and commercial areas through the private provision of road infrastructure, energy, water and sanitation services, as well as mitigation for urban and environmental impacts and the building of social housing provision within the developments (MINVU 1997). Ten specific areas defined as ZUDCs were set by the PRMS after the proposal of a selected group of developers who privately expressed interest in this scheme. Five of these ten ZUDC areas were located in Chicureo, and four of them were under development in 2015 (see Table 8.1).

The ZUDC large residential enclaves in Chicureo were among the first developments to be subject to the environmental assessment system first enacted in 1997. This system established a measurable sustainability standard of construction and operation, and set environmental mitigation and compensation mechanisms for hydrological management, provision of green space, and in some cases, relocation of endangered species and protection of archaeological sites (COREMA 1999, 2000a, 2000b, 2003). Crucially, as part of the environmental impact assessment systems, developments in the ZUDC areas were required to comply with Santiago's metropolitan air quality plan that included measures such as compensation for road use pollution by estimating the projects' future fleet of private vehicles (CONAMA 1998). These measures prompted the replacement of conventional transport vehicles such as taxis and buses with cleaner ones, the development of additional green areas and reforestation, payment of pollution fees and funding of ecological educational centres (COREMA 1999, 2000a, 2000b, 2003).

Once the Plan Chacabuco was approved, the niche market strategy of the developers became the major force in driving the developments. The developers who were awarded the ZUDC scheme in Chicureo were large property development companies who owned very large estates and were strongly linked to the financial and construction industries (Hidalgo *et al.* 2005). These developers devised a market strategy to maximise the land rent of the newly created urban land, and, instead of developing socially mixed areas, they created exclusive neighbourhoods for upper-middle and high-income groups, expanding the model

Table 8.1 Largest and most elite green enclaves in Chicureo

Planning	*Name*	*Housing price*	*Area (ha)*	
			Total	*Urban*
ZUDC	Ciudad Chicureo	High/upper-middle	4,056	1,062
ZUDC	Valle Santa Elena	Middle	1,950	1,950
ZUDC	El Chamisero	Upper-middle	1,597	542
ZUDC	Santa Filomena	–	1,000	335
ZUDC	Pan de Azúcar	High	794	474
Parcelas	Las Brisas de Chicureo	Very high	530	–
Parcelas	Hacienda Chicureo	Very high	190	–

Sources: COREMA 1999, 2000a, 2000b, 2003; FFV 2012; Bermeo 2013; Patagonland 2014.

of gated communities that was emerging in Santiago (Sabatini and Cáceres 2004). These communities would be serviced by private energy, water and sanitation utilities, and connected to Santiago's centre, business and wealthy districts by high-performance motorways (Hidalgo *et al.* 2007).

These enclaves had sustainability features that followed the requirements of the metropolitan master plan and added new attributes to reinforce both the green character of the developments and their exclusivity, such as central parks, ecological reserves, golf courses, large artificial lagoons and pilot waste recycling systems. As the market became the dominant driver of the enclave developments, greenness and localisation became the main characteristics to differentiate these enclaves from other projects in Santiago. From the start, the developers had a distinctive marketing campaign in national newspapers, housing publications and local billboards to promote the developments and their green attributes, which was often uncritically echoed by the media who reported how 'Chicureo becomes a green city' (El Mercurio 2014). A developer characterised Chicureo as a 'scenic green' (Interview 1, Developer, December 2012[6]) because of its rural or peri-urban localisation, premium sport and recreation infrastructures and a 'gardenified' urbanism. While the original advertisements emphasised 'natural' areas and large lawns, they also introduced new features such as ecologically designed parks with native vegetation species to reduce water consumption in the enclaves and to harmonise with the surrounding landscape, and the use of advanced technologies for energy and water efficiency and waste recycling (Figure 8.1).

Figure 8.1 Ecological landscape design of Montepiedra Park in Ciudad Chicureo enclave (source: C. Illanes, used with permission).

Meanwhile, compliance with the requirements of the Plan Chacabuco was uneven. For example, there was flexibility in substituting compulsory measures with others that were easier to achieve (Interview 14, Developer, January 2013). At times, the authorities were unable to compel developers to include mitigation measures such as rainwater collection (Gutierrez 2010). Indeed, the ZUDCs lacked institutional coherence. A ZUDC governance mechanism emerged that was based on ad hoc and direct negotiations between relatively weak regional authorities and the strong and politically connected property developers (Zegras and Gakenheimer 2000; Poduje and Galetovic 2006). Also, rural landowners and other developers began to develop large condominiums to emulate the ZUDC enclaves and benefit from the increased land price and the new provision of infrastructure, connectivity and services, without being subjected to the requirements of the Plan Chacabuco. These new, large and exclusive parcela enclaves developed alongside the ZUDCs substantially increased the population of Chicureo to nearly 30,000 inhabitants, much higher than the original objectives of the Plan (Valencia 2014).

Despite critiques of the social and environmental performance of the enclaves, the planning tools that regulated the ZUDCs were never amended. Instead, successive planning authorities attempted to replicate those attributes of the Plan Chacabuco they considered successful on a wider metropolitan scale, and in 2003 new tools for conditioned planning development by private developers in other rural and peri-urban areas were established. In 2014, a modified PRMS extended the principles of sustainable urban development to the whole metropolitan area (MINVU 2014). While the Plan Chacabuco did not achieve its original objectives, it created a model of green development that is now practiced throughout Santiago.

Maintenance as reproduction of socio-environmental inequalities

As the bulk of planning and environmental conditions of the ZUDCs was established in the building phase – and the parcela enclaves are mostly deregulated – the operation and maintenance of the green enclaves has been left to the general rule of the law and the private market initiative of developers, utilities and infrastructure providers, residents and owners. In practical terms, the maintenance of the green enclaves is realised through a hierarchical division of labour between different stakeholders. As the enclaves still hold large portions of undeveloped land and house units to be sold, developers show great interest in maintaining the enclave environment, to keep both its construction and operation both functional and attractive. Enclave developers practice a maintenance strategy that combines the exercise of social power with environmental control and marketing. For instance, developers set draconian contractual regulations with the homeowners, in which they secure executive powers for themselves for several years, until an administration committee designated by the owners takes charge of the responsibility to manage and fund the cleaning, maintenance and security of the condominium common areas. Meanwhile, the responsibilities for the houses and individual site maintenance are transferred to the individual owners and are controlled through a system of fees and sanctions as part of the bylaws of a very exclusive enclave (Casas de Hacienda 2007).

The maintenance of their greenness demands a continuous flow of materials and energy to and from the developments. Private utilities own and operate the main potable water supply, sewage treatment, electricity and gas supply to service developers, residents and maintenance companies. The most crucial maintenance problem for the green enclaves is groundwater because developers use groundwater sources for cleaning, to irrigate green areas and sports fields, and to maintain artificial lagoons. The enclaves are built in a semi-arid climate area in the Andean piedmont and foothills that has been subject to an extended period of drought (ONEMI 2008). Yet developers and private water utilities have displayed an effective takeover of the local water sources, particularly by controlling the water rights of the Chicureo aquifer[7] and by building large infrastructures to extract and distribute the water (DGA 2013).

Residents have a rather passive role in the maintenance of the enclaves. Apart from caring for their front gardens and paying management fees for common areas and parks, they may choose to sort their recyclables (glass, metal, paper and electronics). In recent years, the water consumption rate of the enclave householders, boosted by the maintenance needs for residential gardens and pools, has led to the highest rate of water consumption in the entire country. For example, a single enclave (Pan de Azúcar) consumes an average 1,120 litres of water per capita per day, nearly six times the national average (Rivera 2011). While the greenness of the enclaves thrives thanks to resource control and technological advanced networks and infrastructures, nearly 20,000 low-income rural inhabitants in the area suffer from severe water scarcity and depend on emergency schemes for domestic water provision from micro water networks and water trucks (Gobernacion Provincia de Chacabuco 2012).

Developers had to build and maintain a complex hydrological system of canals, creeks and lagoons to manage surface water, in addition to the potable water supply and sewage networks. As the legal mandate set by the environmental assessment resolution is limited to the perimeter of the enclave, urban drainage from the development flows offsite and into inadequate or non-existent public drainage systems, causing floods in winter and exacerbating droughts in the lower points of the watershed in summer (Astudillo 2007; INDH 2011). The control of groundwater by private owners has resulted in a stark contrast between the degraded environmental qualities of the low-income areas of the settlements around the enclaves and the 'green' private enclaves.

Living and perceiving green

Residents of the enclaves have a strong perception that they live in an oasis of greenness within a hostile metropolitan urban space. Almost every interviewee used the adjective 'green' to describe the enclave environment. The idea of enclaves as 'the greenest condominiums' in the Chilean northern desert (Interview 43, Resident, April 2013) was presented as the common sense discourse of enclave residents. More than three-quarters of the residents surveyed stated that nature and environment was a very prominent reason to live in a Chicureo enclave, even more important than reasons such as privacy or security. At the same time, interviewed residents saw this greenness as an advantage in the current housing market, as 'green gives more value to the sector and the properties' (Interview 52, Resident, April 2013).

The everyday life of those who inhabit the enclaves involves multiple restrictions that regulate their social and environmental behaviour. Households are constrained by contracts that establish several punishable covenants to maintain the aesthetic form of the neighbourhood, the programmed functions of the spaces and infrastructures, and a strict code of behaviour to maintain the 'right' environment of the green enclaves (Interview 42, Resident, April 2013). At the same time, residents follow a social code of conduct driven by their own group of reference and status position, which may establish that having a solar heating on the roof is aesthetically unacceptable or that car sharing with neighbours is socially undesirable (Interview 23, Resident, March 2013). Workers are also subjected to regulations and social codes. For instance, housemaids are not allowed to walk in the streets and are required to wear uniforms. Meanwhile, childcare providers are banned from exclusive areas of the enclave golf clubs, which has already caused public national outrage (Publimetro 2012). The marketing of the enclaves has increasingly emphasised resource efficiency alongside more traditional green attributes such as rural localisation and landscape. This feeds into a more active environmental role for residents that emphasises consumption behaviour and regulation of the devices or services that they should buy.

The greenness of the enclaves and their sustainability are perceived as embedded either in the enclaves design or built environment. The discourses of the residents also reveal a self-perception as a group of having both low ecological consciousness and low engagement with actions for environmental improvement (Interview 49, Resident, April 2013). Enclave residents interviewed showed a lack of awareness of their own environmental footprint. They are unaware, for example, that their footprint is larger than average due to their use of private motorised vehicles and their high rates of energy and water consumption. For instance, some enclave residents considered their water consumption was average or even 'miserably low' compared with the average Chilean (Interview 42, Resident, April 2013) and they did not recognise any responsibility for reducing their resource consumption or considering lifestyle alternatives. Those who did mention the need for change in environmental behaviour thought that ecological consciousness should emerge as a result of public authority action (Interview 45, Resident, April 2013) or as a result of the adoption of individual economic rationality to address environmental problems. As one resident stated, 'what creates [ecological] consciousness is one's own wasting of resources' (Interview 31, Resident, March 2013).

The residents perceived their possibilities for action in exclusively individualist terms. In the interviews, collective action was never mentioned as a legitimate mechanism to address resource-dependence issues nor was the resident owners' association cited as a legitimate institution to discuss and promote pro-sustainability actions and behavioural change. In the few cases in which residents have engaged in collective action, this has always been directed toward the protection of 'their environment' threatened by public projects, such as in protests against the construction of a sewage treatment plant in 2009 or against the extension of the electricity network to an adjacent enclave in 2014 (No a la Farfana de Colina 2010; Pulso 2013).

Conclusions

A decade and a half since being enacted, the policy innovations that led to the development of the experimental ZUDC enclaves in the rural areas of metropolitan Santiago can be considered successful in urbanising Chicureo, but the advancements in social, economic and ecological sustainability of these experimental enclaves is disputable. Although the development process followed policy guidelines, the resulting enclaves have not achieved many of the desired aims. At the same time, the development of Chicureo as a housing market product under a 'green' niche strategy proved highly successful, boosting the development and population of enclaves with green attributes based on market-led ecological design, systems efficiency and technological innovations.

The analysis of how experiments are maintained and inhabited adds a layer of complexity by emphasising not only the extent to which the state is implicated in reproducing the structures for the functioning of the market, but also in the constitution of subjects through the deployment of a will to improve (Bulkeley *et al.* 2014). These experiments are not just directed towards the regulation of behaviour but also towards the regulation of the particular ecologies that sustain these developments. The analysis indicates that the specific modes of environmental maintenance undertaken in the enclaves are used to keep their greenness marketable. They play a crucial role in transferring environmental costs to other areas and social groups. In this way, they produce and reproduce environmental injustices by creating a socially segregated area of affluent populations with premium infrastructures and services and concentrating crucial recourses as water and land in a few private hands. Beyond the stark aesthetic greenness of the enclaves, parameters of urban sustainability defined by the state's mitigation measures and the developers' marketing reveal that there is little connection between actions that are emphasised as being green and their actual impact on the environment. More worryingly, the interviews reveal a distorted perception about the extent to which the enclaves are sustainable and why and who should bear the responsibility for disproportionate levels of resource consumption.

The case of Chicureo demonstrates the emergence of market-led forms of environmental governance. Not only are local populations and the government largely absent from the enclave's maintenance but, also, residents play a rather passive role in the maintenance of the enclaves, except for their individualistic role as consumers to exert a right to choose a ready-made green product in the market. Residents adhere to a market environmentalist ideology whereby they believe they have bought a product in which the guarantee of greenness is already embedded.

What conclusions can we draw in relation to the encounter between the governance of socio-ecologies through experimentation and the neoliberal regulation of the territory? The case is bleak in that it demonstrates how the neoliberal context shapes both experimentation possibilities – the assemblage or making of the experiment – and the way it is deployed in a spatial context (how it is maintained and lived). In doing so, the experiment resonates with the analysis of While and colleagues (2010) about the limits of progressive environmental action. It also shows that experiments themselves may serve to justify and promote an agenda of neoliberalisation whereby enclave-based urbanisation is predicated.

In a wider perspective, the enclave experiment seems to align green concerns with the requirements of the neoliberal project rather than contest it, but it is difficult to consider this as a mere reproduction of a stable and dominant regime. Instead, we suggest that green enclaves may be understood as an expression of a struggle to challenge neoliberal urbanisation, and in the process, situate urban sustainability at the centre of the scenario for establishing a new regulation of urbanisation and the environment. The disjunctive duality in this case is not between statism and the previous dictatorial laissez faire environmentalism but between the specific modes of constitution of a socio-nature that could drive a truly egalitarian democracy and those that would reinforce exclusion and inequality.

Notes

1 The interviews were conducted with 19 residents, 11 state officials, 11 consultants, 7 activists, 6 scholars, 5 developers and 2 workers. Some interviews addressed more than one interviewee.

2 Residents in the ZUDC enclaves were surveyed using the online provider Survey Monkey. The survey was conducted for one month starting on 15 April 2013.

3 The weekly magazine is *Vivienda y Decoración* (*Housing and Decoration*), and the sample consisted of 47 issues from December 2012 to December 2013.

4 From December 2012 to May 2013.

5 This translates literally as 'enjoyment allotment'. The legal basis for this subdivision was a 1980 military decree established to facilitate rural inheritances and promote the colonisation of the far south of the country. Most of the subdivisions were built in the northern province of Chacabuco. In the area of Chicureo, nearly 50,000 hectares were legally subdivided into ½ hectare plots (see Naranjo 2009).

6 Interviewees' names are anonymised to comply with research ethics standards.

7 Under Chilean law, property owners are entitled to free water rights from surface and subsurface sources by request when available. This right is then market tradable. In the case of Chicureo, both developers and water utilities were granted water rights until the aquifer was declared saturated by the water authority (Budds 2004; ONEMI 2008).

References

Astudillo, B. (2007). *Análisis de Técnicas Alternativas y Complementarias para las Soluciones de Aguas Lluvia en una Cuenca. Aplicación al Valle de Chicureo*. Tesis de Grado, Universidad de Chile. [Online]. Available: http://www.tesis.uchile.cl/tesis/uchile/2007/astudillo_b/sources/astudillo_b.pdf [16 October 2013].

Bermeo, M. (2013). Modificaciones al plano regulador metropolitan afectarían a desarrollos de gran escala de Pudahuel y Lampa. *Diario Financiero*, 20 November 2013.

Budds, J. (2004). Power, nature and neoliberalism: the political ecology of water in Chile. *Singapore Journal of Tropical Geography*, 25: 322–342.

Bulkeley, H., Castàn Broto, V. and Edwards, G. (2014). *An Urban Politics of Climate Change: Experimentation and the Governing of Socio-Technical Transitions*. London: Routledge.

Caldeira, T. (2000). *City of Walls: Crime, Segregation, and Citizenship in São Paulo*. Berkeley: University of California Press.

Casas de Hacienda (2007). *Reglamento de Co-Propiedad*. [Online]. Available: http://www.piedraroja.cl/descargas/2.Reglamento_Copropiedad_Casas_de_Hacienda.pdf [15 September 2014].

Castán Broto, V. (2015). Contradiction, intervention, and urban low carbon transitions. *Environment and Planning D: Society and Space*, 33: 460–476.

CONAMA (1998). *Plan de Prevención y Descontaminación Atmosférica de la Región Metropolitana*. Comisión Nacional del Medio Ambiente. [Online]. Available: http://www.leychile.cl/Navegar?idNorma=121128&tipoVersion=0 [3 May 2014].

COREMA (1999). *Resolución Calificación Ambiental DIA Proyecto Pan de Azúcar*. [Online]. Available: http://seia.sea.gob.cl/externos/admin_seia_web/archivos/1579_19991216_RE.doc [12 February 2013].

COREMA (2000a). *Resolución Calificación Ambiental DIA Proyecto Chicureo Ciudad*. [Online]. Available: http://seia.sea.gob.cl/externos/admin_seia_web/archivos/2474_2000_8_10_RE.doc [12 February 2013].

COREMA (2000b). *Resolución Calificación Ambiental DIA Proyecto El Chamisero*. [Online]. Available: http://seia.sea.gob.cl/externos/admin_seia_web/archivos/2172_20000914_RE.doc [12 February 2013].

COREMA (2003). *Resolución Calificación Ambiental DIA Proyecto Valle Santa Elena*. [Online]. Available: http://seia.sea.gob.cl/externos/admin_seia_web/archivos/5699_2003_4_17_RE.doc [12 February 2013].

Coutard, O. and S. Guy (2007). STS and the city: politics and practices of hope. *Science, Technology and Human Values*, 32: 713–734.

Davis, M. (1998) *Ecology of Fear: Los Angeles and the Imagination of Disaster*. New York: Vintage Books.

DGA (2013). *Derecho de Agua Concedidos Region Metropolitana*. Direccion General de Aguas. [Online]. Available: www.dga.cl/DGADocumentos/Derechos_Concedidos_XIII_Region.xls [27 February 2013].

El Mercurio (2014). Chicureo se transforma en una ciudad verde. *El Mercurio*, 25 July 2014. [Online]. Available: http://www.edicionesespeciales.elmercurio.com/destacadas/detalle/index.asp?idnoticia=201407251652285&idcuerpo=1274# [29 August 2014].

Evans, J. and Karvonen, A. (2014). 'Give me a laboratory and I will lower your carbon footprint!' – Urban laboratories and the governance of low-carbon futures. *International Journal of Urban and Regional Research*, 38: 413–430.

FFV (2012). *Proyectos en Venta: Las Brisas de Chicureo*. [Online]. Available: http://www.ffv.cl/las-brisas-de-chicureo.html [18 March 2013].

Gieryn, T. (1999). *Cultural Boundaries of Science*. Chicago: University of Chicago Press.

Gobernacion Provincia de Chacabuco (2012). *Proyectos*. [Online]. Available: http://www.gobernacionchacabuco.gov.cl/info_inversiones.html [14 January 2013].

Gutiérrez, M. (2010). MOP exige a inmobiliarias millonarias obras de aguas lluvias en Chicureo. *El Mercurio*, 3 June 2010. [Online]. Available: http://chicureo.com/cronica/noticias_principales/2010/10_1.shtml [12 March 2013].

Hackworth, J. (2007). *The Neoliberal City: Governance, Ideology, and Development in American Urbanism*. Ithaca, NY: Cornell University Press.

Hidalgo, R., Salazar, A., Lazcano, R., Roa, F., Alvarez, L. and Calderón, M. (2005). Transformaciones socioterritoriales asociadas a proyectos residenciales de condominios en comunas de la periferia del Área Metropolitana de Santiago. *Revista INVI*, 20(54).

Hidalgo, R., Borsdorf, A. and Sánchez, R. (2007). La expansión residencial amurallada en la reconfiguración metropolitana en Santiago de Chile. Santiago de Chile: Comisión Económica para América Latina y el Caribe, CELADE.

Hodson, M. and Marvin, S. (2010). Can cities shape socio-technical transitions and how would we know if they were? *Research Policy*, 39: 477–485.

INDH (2011). *Desastre Ecológico en Humedal de Batuco*. Instituto Nacional de Derechos Humanos, Mapa de Conflictos. [Online]. Available: http://www.indh.cl/mapaconflictos/conflicto/detalle/76 [17 January 2014].

Jonas, A. E., Gibbs, D. and While, A. (2011). The new urban politics as a politics of carbon control. *Urban Studies*, 48: 2537–2554.

Karvonen, A. and van Heur, B. (2014). Urban laboratories: experiments in reworking cities. *International Journal of Urban and Regional Research*, 38: 379–392.

Markard, J., Raven, R. and Truffer, B. (2012). Sustainability transitions: an emerging field of research and its prospects. *Research Policy*, 41: 955–967.

MINVU (1997). *Plan Regulador Metropolitano de Santiago: Provincia Chacabuco*. [Online]. Available: http://www.seremi13minvu.cl/opensite_20080710120510.aspx [17 January 2014].

MINVU (2014). *Modificación del Plan Regulador Metropolitano de Santiago: Actualización, Extensión y Reconversión*. [Online]. Available: http://www.iconstruccion.cl/wp-content/uploads/2014/04/PRMS-100-2014.pdf [25 July 2014].

Naranjo, G. (2009). El rol de la ciudad infiltrada en la reconfiguración de la periferia metropolitana de Santiago de Chile. *Estudios Geográficos*, 70: 205–229.

No a la Farfana de Colina (2010). *Lo Logramos: Corema Rechaza el Proyecto PTAS de Colina en Forma Unánime*. [Online]. Available: http://vivoenlopinto.wordpress.com/2010/02/25/lo-logramos-corema-rechaza-el-proyecto-ptas-de-colina-en-forma-unanime-20-votos-en-contra-0-a-favor/ [11 October 2014].

ONEMI (2008). *Informe Actualizado Sequia*. Gobierno de Chile, Oficina Nacional de Emergencia, Ministerio del Interior. [Online]. Available: http://repositoriodigitalonemi.cl/web/bitstream/handle/123456789/402/informe%20Sequia%202008%20actualizado.pdf?sequence=1 [8 October 2014].

Patagonland (2014) Cifras destacadas. [Online]. Available: http://www.patagonland.cl/proyectos/mirador-pie-andino [13 March 2014].

Petermann, A. (2006). ¿Quién extendió Santiago? Una breve historia del límite urbano, 1953-1994. In Galetovic, A. (ed.), *Santiago. Dónde estamos y hacia dónde vamos*. Santiago, Chile, Centro de Estudios Públicos (CEP), 205–230.

Poduje, I. and Galetovic, A. (2006). El globo y el acordeón: planificación urbana en Santiago, 1960–2004. In Galetovic, A. (ed.), *Santiago. Dónde estamos y hacia dónde vamos*. Santiago, Chile, Centro de Estudios Públicos, 131–176.

Publimetro (2012). Chicureo: empresario que denunció discriminación dona terreno a su nana. *Diario Publimetro*, 13 January 2012. [Online]. Available: http://www.publimetro.cl/nota/cronica/chicureo-empresario-que-denuncio-discriminacion-dona-terreno-a-su-nana/xIQlan!fPrkA0dxlpE/ [29 January 2013].

Pulso (2013). Vecinos de Chicureo exigen a Chilectra trazado subterráneo para instalar torres. *Diario Pulso*, 11 December 2013. [Online]. Available: http://www.pulso.cl/noticia/empresa-mercado/empresa/2013/12/11-35320-9-vecinos-de-chicureo-exigen-a-chilectra-trazado-subterraneo-para-instalar-torres.shtml [12 October 2014].

Rivera, R (2011). Sectores Altos de Santiago son los que Tienen el mayor consumo de agua en Chile. *Diario EMOL*, 21 August 2011. [Online]. Available: http://www.emol.com/noticias/economia/2011/11/24/514225/np-fds-sobre-los-1000-litros-de-agua-por-persona-son-consumidos-por-habitantes-de-sectores-altos-del-pais.html [9 October 2014].

Sabatini, F. and Cáceres, G. (eds) (2004). *Barrios Cerrados en Santiago de Chile: Entre la Exclusión y la Integración Residencial.* Santiago de Chile: Lincoln Institute of Land Policy/Instituto de Geografía, P. Universidad Católica de Chile.

Schipper, S. (2014). The financial crisis and the hegemony of urban neoliberalism: lessons from Frankfurt am Main. *International Journal of Urban and Regional Research*, 38: 236–255.

Swyngedouw, E. (1997). Power, nature, and the city. The conquest of water and the political ecology of urbanization in Guayaquil, Ecuador: 1880–1990. *Environment and Planning A*, 29: 311–332.

Swyngedouw, E. (2004). *Social Power and the Urbanization of Water: Flows of Power.* New York: Oxford University Press.

Valencia, M. (2014). Chicureo alista proyectos para recibir a 120 mil habitantes más en la próxima década. *El Mercurio*, 26 March 2013. [Online] Available: http://impresa.elmercurio.com/Pages/NewsDetail.aspx?dt=2014-03-26&dtB=26-03-2014%20 0:00:00&PaginaId=11&bodyid=3 [12 April 2014].

While, A., Jonas, A. and Gibbs, E. (2010). From sustainable development to carbon control: eco-state restructuring and the politics of urban and regional development. *Transactions of the Institute of British Geographers*, 35: 76–93.

Zegras, C., and Gakenheimer, R. (2000). *Urban Growth Management for Mobility: the Case of the Santiago, Chile Metropolitan Region.* Report. Boston, MA: Lincoln Institute of Land Policy and MIT Cooperative Mobility Program.

9 Urban mobility experiments in India and Thailand

Duke Ghosh, Frans Sengers, Anna J. Wieczorek, Bipashyee Ghosh, Joyashree Roy and Rob Raven

Introduction

Asia's rapidly growing cities are gearing up to meet increasing mobility needs while simultaneously striving to achieve sustainability goals. A number of new innovations are being introduced in the form of experiments aiming to change the systems and rule-sets that currently dominate the provision of mobility services in Asian cities. This chapter analyses selected cases of mobility experiments in India and Thailand and explores the strategies that the niche actors deploy to navigate the challenges posed by incumbent socio-technical regimes. The chapter concludes that whilst niche actors tried to stretch-and-transform technological, infrastructural and cultural dimensions of regimes, they opted to fit-and-conform with public policy and political power dimensions.

Many cities in Asia are witnessing an urgency to reorganize, reconstruct and reorient themselves in response to rapid urbanization (United Nations 2014). Simultaneously, they are striving to conform to the goals of sustainability and liveability (Hildebrand *et al.* 2013) as 'the battle for sustainability will be won or lost in cities' (Bai *et al.* 2010: 312). It is widely recognized that experimentation with new socio-technical configurations – whether in the form of 'urban climate change experiments' (Bulkeley *et al.* 2015), 'urban labs' (Karvonen and van Heur 2014) or 'niche experiments' (Hoogma *et al.* 2002) – is a promising but challenging way to bring about sustainability transitions in cities in a variety of societal domains.

As both a driver and an outcome of rapid growth, mobility is an integral part of urban development and one key area where environmental sustainability is proving to be a major challenge (Moriarty and Honnery 2008; Banister 2011; GEA 2012; Geels *et al.* 2012; IPCC 2014). Many Asian cities are responding to this challenge by experimenting with new transport technologies and novel mobility practices that embody the promise of sustainability gains. Since these innovative socio-technical configurations seek to undermine some of the established rules of the game, their introduction implies a struggle with incumbent actors and routines. In the context of this struggle, an important research question is: how do the actors involved in these experiments navigate the tensions that arise when introducing such innovations?

In this chapter, we explore four selected cases in cities in India and Thailand to tease out some of the interesting features of the navigational strategies followed by experimental actors. The chapter is built around three parts. First, we discuss the theoretical framework and the research methodology. Next, we delve into the empirical cases to demonstrate the dynamics by which the actors negotiate the pressures from the incumbent regime. And in conclusion, we summarize our main argument and consider the patterns that emerge from our cases.

Empowerment and niche strategies

The 'sustainability transitions' literature argues that achieving sustainability requires a systemic and radical change in the way human needs are being satisfied (Markard *et al.* 2012). Because of their path-dependent character, these systems are often referred to as 'socio-technical regimes' that embody the 'rules of the game' that structure the provision of particular societal needs such as mobility. Path-breaking innovations offering alternative ways of meeting the needs frequently emerge outside of the regime; that is, in 'niches' that act as protective spaces, where experiments with novel socio-technical configurations can develop relatively free of the full brunt of mainstream market selection. A system transformation occurs when radical novelties are sufficiently developed and when the 'landscape', defined as a broad exogenous environment, exercises sufficient pressure on the prevailing regimes, alters them and makes them unstable (Geels 2002).

The process of regime change, however, is not easy because regimes are stabilized on a number of dimensions. Smith and Raven (2012, based on Geels 2002), depict six such dimensions. The first dimension relates to existing industry structures that encompass established network relations, user–producer interactions, shared routines and heuristics, existing capabilities and resource allocation procedures. The second dimension concerns dominant technologies and infrastructures that are supported by articulated technical standards and infrastructural arrangements. Third is the established knowledge base with its guiding principles and socio-cognitive processes. The fourth covers markets and dominant user practices stabilized by market institutions, supply and demand, price mechanisms, user preferences and routines. The fifth includes status quo public policies and political power stabilized by means of the prevailing regulations, policy networks and relations with incumbent industries. Finally, the sixth dimension considers cultural and symbolic meanings of technology and innovation. As such, socio-technical regimes may lead to premature rejection of alternative novelties, because their costs are not represented in end-user prices, because they require inconvenient user practices, because insufficient resources are attributed to new knowledge development, because they do not fit with existing industry structures, and so on.

To deal with the stability of the regimes and to link to the wider processes of social change, niche actors deploy various strategies. Smith and Raven (2012) identify two different strategies of empowerment. The first strategy involves processes that make niche innovations competitive within unchanged selection environments. This is referred to as a 'fit-and-conform' strategy. In this case, niche protection is no longer necessary and can be removed because the innovation is 'empowered' and its growing competitiveness enables its widespread diffusion. The niche innovation is developed in a way that it fits into and conforms to a relatively unchanged selection environment. The second empowerment strategy can also occur when the niche innovation contributes to changes in mainstream selection environments in ways favourable to that niche innovation (a 'stretch-and-transform' strategy). In this case, some of the niche practices and features are institutionalized as new norms and routines in a transformed regime. Because of stretching and transforming regimes, this strategy is not only dependent on internal niche dynamics but also on the external processes of regime destabilization and changes in the broader landscape context.

To understand the empowerment strategies in our case studies, we take a navigational approach based on following the niche actors. This diverges from the dominant approach in Transition Management that is based on defining the regime from the outset of the study (see, for example, Verbong *et al.* 2008). By following the niche actors, we can learn about the challenges these actors face in the process of system change, whether through fit-and-conform or stretch-and-transform strategies. In order to understand where the critical tensions and struggles occur, we analyse the regime dimensions – industry structure, technology and infrastructure, knowledge base, public policies and political power, users' relations and markets, and cultural significance and associations of the regime – and identify the type of strategy that the niche actors deploy (fit-and-conform or stretch-and-transform) to link to the wider processes of social change.

We apply this approach to four cases: electric rickshaws in New Delhi, motorcycle taxi-meters in Bangkok and Bus Rapid Transit (BRT) systems in Ahmedabad and Bangkok. The cases form part of a larger research project carried out in the context of the NWO-WOTRO Integrated Programme on 'Experimenting for Sustainability in India and Thailand: A transitions perspective on sustainable electricity and mobility initiatives' (see Ghosh *et al.* 2013; Sengers and Raven 2014, 2015). In this chapter, we highlight the dimensions of 'technology and infrastructure', 'cultural significance' and 'public policy and political power'. These dimensions reveal important controversies and navigational issues in our cases and highlight elements that have been mentioned as lacking in other accounts of niche experiments and mobility transitions, notably politics/power (Smith and Raven 2012; Tyfield 2014) and culture (Sheller 2012; Valderrama and Vogel 2014).

The primary means of data collection for this research included semi-structured and exploratory interviews with a range of actors involved in the experiments. Consultations with technology developers and implementers revealed the articulations, motivations and goals of the experiments, as well as the struggles and tensions they confronted. Interactions with policy makers and governments highlighted the dominant paradigms of the existing regimes. Interviews with users and, in some cases, members of civil society helped to reveal a range of socio-cultural perspectives. In addition, we collected secondary data from grey literature.

Electric rickshaws in New Delhi

In Indian cities, cycle rickshaws occupy a significant position. As a form of non-motorized intermediate means of transport (IMT) they carry both people and goods across short distances.[1] The mode provides employment to millions of poor and unskilled people (Rajvanshi 2002).[2] With little improvement in design, cycle rickshaws have been described as being ergonomically inefficient (Rajvanshi 2002), causing chronic health problems to the drivers (McMichael 2000).[3] Diseases including tuberculosis, physical weakness, and so on, are common amongst the drivers (Begum and Sen 2004; Maji *et al.* 2010; Khan *et al.* 2012).

Scientists and planners in India have long been trying to re-engineer and redesign cycle rickshaws to improve the drivetrain and reduce stress on the drivers (Maji *et al.* 2010). These efforts resulted in the development of electric rickshaws (e-rickshaws) driven by a mix of motor and manual power. E-rickshaws were launched in New Delhi in 2008 (Figure 9.1).[4] In 2010, the Municipal Corporation of Delhi (MCD) decided to replace the fleet of cycle rickshaws with e-rickshaws. This decision resulted in the partial replacement of conventional cycle rickshaws as well as auto rickshaws. In 2013/2014, approximately 100,000 e-rickshaws operated in New Delhi (Chakravarty 2014). Their introduction influenced the fleet mix in the IMT space in the city but because of the incompatibility of the e-rickshaws with the existing regime dimensions, a number of conflicts emerged. Beyond the various techno-economic conflicts that the actors associated with the experiment had to address, cultural and policy dimensions merit specific attention.

Figure 9.1 E-rickshaws on the streets of New Delhi.

Describing cycle rickshaws as slower and inhumane and auto rickshaws as polluting was a major strategy to garner public support, the e-rickshaw was proclaimed as a 'clean pedicab' (CMERI 2010). Simultaneously, improved income opportunities were promised for the drivers of e-rickshaws, through higher speed and thus, more passenger kilometres (Maji *et al.* 2010). This 'stretch-and-transform' strategy linked e-rickshaws to the sustainability paradigm and presented them as having the potential for considerable sustainability gains (Maji *et al.* 2010) – including health and income benefits and reduction of emission and pollution.

An interesting interplay between the state institutions, political actors and technology developers was evident. E-rickshaw manufacturers designed the technology following Central Motor Vehicles Rules (CMVR).[5] The rules carry a provision whereby vehicles with motor power of less than 250 watts are designated as non-motorized vehicles (NMV). Governed by the rules of the municipal authorities, NMVs enjoy a special legal space.[6] They are not required to have insurance and/or a licence for a driver. The technology developers paid close attention to this provision of the law and restricted the motor power of the vehicles to within this limit. The strategy reduced the net cost of ownership of the vehicle and enabled anyone to drive an e-rickshaw. The strategy helped city authorities to negotiate with the unions controlling cycle rickshaws to substitute cycle rickshaws with e-rickshaws. The MCD permitted e-rickshaws to ply anywhere in New Delhi (Chakravarty 2014), as no new or additional road infrastructure was required for these vehicles. Cycle rickshaw drivers saw this as an opportunity for accessing larger service areas. As a result, an increasing number of e-rickshaws rapidly filled the streets of New Delhi.

However, within a short time, the e-rickshaws attracted controversy by causing a number of road accidents (TIO 2014) attributed to untrained and unlicensed drivers. Further, the absence of insurance deprived the victims of compensatory benefits. Additionally, some studies revealed that many of the vehicles exceeded the stipulated motor power (Delhi High Court 2014). This initiated a debate concerning whether e-rickshaws needed to be regulated by the CMVR. In April 2014, the Central Ministry for Road Transport imposed a temporary ban on e-rickshaws in New Delhi. Subsequently, the court declared the plying of electric rickshaws in the city as 'illegal' until appropriate legal provisions are in force, citing 'prima facie they are a hazard to other traffic as well as citizens' (Delhi High Court 2014: 25). The issue is 'sub judice'. This case suggests that the actors navigated through the cultural significance of the regime through 'stretch-and-transform' strategies. They also adopted 'fit-and-conform' strategies to negotiate the legal and policy space of the existing regime. However, the latter strategy backfired as larger legal and social issues were not adequately addressed.

Motorcycle taxi-meters in Bangkok

In 2010, two entrepreneurs developed a small electronic gadget to be used as a taxi-meter. As opposed to other taxi-meters installed in air-conditioned cabs, this portable device was designed specifically for mounting on the handlebars of motorbikes (Figure 9.2). Because there is a great, and growing, number of motorcycle taxis in cities throughout the developing world (Gwilliam 2002; Kumar 2011) – all of which operate without meters[7]– the entrepreneurs thought that there might be a lucrative market for such a product. However, before any promise of a potential multi-million dollar industry for metered motorbike mobility could be realized, the newly designed device would first have to be trailed and tested in 'real-life' conditions. Because they were based in Thailand, the two entrepreneurs decided to try and set up a small-scale pilot project in Bangkok.

As part of the feeder system essential to the city's mobility, 200,000 motorcycle taxis utilize Bangkok's roads and alleyways on a daily basis. Although over half of the drivers are formally registered, most are nonetheless forced to pay bribes to locally powerful people (army, police, politicians, and so on) who control a particular area. In recent years, however, a number of high-level politicians (most notably ex-prime minister Thaksin Shinawatra), as well as a new union-like association of politically active motorcycle taxi drivers, mobilized to rid the sector of this kind of institutionalized corruption (or 'dark influence' as it was called in a government campaign). One of the key challenges for setting up an experiment with the new metering device would revolve around how to operate within this murky socio-political environment. In other words, how would the entrepreneurs navigate the regime dimension of public policy and political power?

To start this process of navigation, the two entrepreneurs devised two opposing strategies for the experimental introduction of their taxi-meter. The first potential

Figure 9.2 Motorcycle taxi-meter trial in Bangkok.

strategy would be to 'inject a virus into the system' that involved the organic, bottom-up introduction of their device by approaching drivers directly. They reasoned that 'buying out' a queue and mounting the device on a few motorcycles might give these drivers an edge over their competitors and that their metering gadget might spread like wildfire. Confronted with this new phenomenon, the authorities would eventually need to come to terms with the changed reality and implement some kind of meaningful regulation to address this new technology. The second potential strategy would be to cooperate with government officials from the start. In this scenario, the entrepreneurs would first try to find willing partners inside the transport authority and negotiate on the possibilities for experimentation, possibly setting up a formal pilot project. The advantage here would be that having 'done the dance' with government agencies from an early stage would pay off in the long run, through the establishment of a legal fare rate and meters as a legal requirement for all motorcycle taxis.

The entrepreneurs opted for the second strategy, aware that the success of the experiment would hinge on the actions of a bureaucratic system mired in stifling polarization and political crisis. Even if interested officials could be found within the plethora of transport agencies with unclear and overlapping mandates – each with their own interests, affiliations and factions – there would be no guarantee that they would want to work together with the entrepreneurs and, indeed, with each other.

In late 2010, the entrepreneurs – aided by a few charismatic friends whom they perceived as skilled in the art of political navigation – approached government officials. After the deputy governor responded favourably, a public meeting was held by representatives of the governor, the Ministry of Transport, the Treasury Department and district chiefs of police. All parties pledged their support in setting up an experiment. In the process of negotiating the details of the upcoming trial, some of the authorities suggested that a massive demonstration project be conducted by supplying meters to all the motorcycle taxi queues around Victory Monument, one of the city's busiest traffic intersections, with the media present in full force at the launch. The entrepreneurs rejected these suggestions, because they feared that some actors might then publicly oppose the device for political gain. They would rather conduct a small-scale trial elsewhere with less at stake.

In late 2011, a small-scale trial was started in another part of town with only 30 devices and with the support of local police chiefs and two groups of drivers. Up to this point, the union-like motorcycle taxi association had been excluded because many of the involved officials did not view it as a legitimate stakeholder. One week into the trial, however, the entrepreneurs sent an envoy to the association headquarters armed with three things: a device (to demonstrate how the taximeter worked), an iPad (to show a movie clip of the experiment), and a bouquet of red roses. The colour red was not a coincidence since the entrepreneurs viewed the association as supporters of ex-prime minister Thaksin and the red-shirt movement established in his tracks. As a gesture of approval, the association presented the entrepreneurs with their association flag.

As such, the 'going to government' strategy combined with the appeasement of the motorcycle taxi union was not geared to directly challenge power relations. Rather than seeking to 'stretch-and-transform' the motorcycle taxi system, it points to a cautious approach geared to 'fit-and-conform' to political power as a way of navigating the socio-political context. Reflecting on this process of political navigation in setting up the experiment, one of the entrepreneurs recounted: 'of course you never know what's really going on in the inside … but at least those people in power didn't look at [the device] as a threat'. In the eyes of the entrepreneurs, however, equipping motorcycle taxis with their device had broader implications. In their view, the device was more than a meter; it was a platform upon which additional functionalities could be built, such as a black-box recorder to be called on in case of an accident and a GPS tracker to realize localized advertising to passengers. As a technology of surveillance, such a device might reshape power relations to the detriment of the drivers, but as a technology of mediation it might work to their benefit. The mere fact of having a motorbike with a meter conveys the image of a 'bona fide' taxi driver, which could be instrumental in legitimizing the motorcycle taxi profession. Indirectly, we might speculate, if motorcycle taxi drivers perceive this as a way to empower themselves and their profession, it could possibly even help to undermine the reproduction of certain informal institutions such as paying informal site rent and the associated chain of privilege and corruption. Argued along these speculative lines, the entrepreneurs represent their device as a 'mediating technology (Furlong 2010: 1) – a small piece of additive technology, which might be capable of bringing about big changes by shifting long-accepted socio-technical relationships in cities. As such, it represents a potential Trojan horse geared to 'fit and – eventually – transform' public policies and political power.

Bus Rapid Transit (BRT) in Ahmedabad

Like most Indian cities, Ahmedabad's mobility regime is marked with the coexistence of various modes of transport – motorized and non-motorized, public and private. Ahmedabad Municipal Transport Services (AMTS) is a major actor in the provision of public buses, a service that has deteriorated over the years (Mahadevia *et al.* 2013). This made the public transport regime in Ahmedabad unstable. Rapid economic growth of the state had also contributed to a gradual modal shift towards motorized and private transport. The shift resulted in traffic congestion, air and noise pollution, and other negative impacts.

It was against this backdrop that the JANMARG BRT project was launched in 2006. The goal was to meet the demand for faster and reliable public transport to provide access to the expanding limits of the city, while simultaneously addressing congestion and environmental problems. Operational since 2009, the JANMARG BRT consists of a fleet of sophisticated buses that runs along dedicated corridors (Figure 9.3). The project was supported by India's national government through the Jawaharlal Nehru National Urban Renewal Mission (JnNURM) and

National Urban Transport Policy (NUTP). However, in its attempts to displace private vehicles, this experiment faced challenges in coexisting with and competing against the private transport regime. In discussing some of the strategies through which the actors navigated these challenges, we focus on the regime dimensions of technology and infrastructure, and cultural significance.

The existing regime infrastructure had already been proven inadequate to handle the traffic of the city. Using the existing infrastructure for the BRT experiment would have meant compromising on the speed and time of public transit while further aggravating congestion. Thus, the BRTs would become a source of irritation to the citizens of Ahmedabad. To overcome the challenges posed by the incumbent infrastructure, the actors decided to create dedicated BRT corridors. This approach allocated a separate space for the BRTs and ensured that the existing traffic remained unaffected. Together, based on the results of user surveys, the actors accorded importance to the provisioning for IMT and non-motorized transport that would act as 'feeders' to the BRT. Dedicated footpaths and cycle lanes were integrated in the infrastructure design for the BRT. As such, a potential co-benefit of the project involved more facilities for walking and cycling, even amongst non-BRT users. The actors asserted that the experiment not only created new infrastructure but fostered a systemic change in the overall mobility infrastructure of the city.

The experiment was positioned as a constellation of a set of new and novel technologies – markedly different from the technologies existing in the incumbent system. Buses with improved fuel efficiency (promising reduced emission per unit of distance travelled) and exhaust (promising reduced pollution), and an intelligent

Figure 9.3 BRTs in Ahmedabad.

tracking system (promising reliability), were introduced. With the systemic change in the infrastructure and deployment of novel technologies, the actors described the experiment as a 'sustainable' and 'reliable' solution to the anticipated increase in future transport demand. The actors successfully communicated the promises of sustainability gains accruing from the experiment, together with the shortcomings of the regime in harnessing financial resources required for the experiment. Notably, the articulations and communications were based on a high degree of background research and knowledge support by academic institutions and policy research organizations to underpin their claims.

Further, the actors stated clearly that the experiment was not intended to replace the different existing modes of transport. Rather, the BRT would stretch the limits of the incumbent mobility infrastructure while ensuring that the experiment also fitted comfortably into the regime by complementing other modes of transport. The technology strategy was 'stretch-and-transform', while the cultural strategy was 'fit-and-conform', promising a comprehensive and sustainable transport network in the city of Ahmedabad.

In developing economies, vehicle ownership shares a positive relation with rising economic affluence and social status (Gakenheimer 1999). Against this cultural backdrop, the experiment in Ahmedabad needed to increase acceptability of the public BRT and induce drivers to shift away from private vehicles. Simultaneously, the experiment had to struggle also against the cultural bias of buses as a service for low-income residents. Further, when many other cities in India (e.g. Kolkata, New Delhi, Mumbai) were expanding or implementing metro rail, pitching for the BRT was a challenge. Strategies were designed to promote a positive image of buses and make the system acceptable to the public. By nature, these strategies were 'stretch-and-transform'.

Naming the system 'JANMARG', meaning 'people's pathway' was a device to signify 'inclusivity'. Further, at the inception stage, the transit system was framed as connecting to the need for an 'international look of Ahmedabad' (Mahadevia *et al.* 2013).[8] The United Nations highlighted this project to demonstrate how addressing climate change is not a burden (Goswami 2012) and this endorsement was advertised heavily to enhance the image of the BRT. Buses plying in the BRT are designed for improved comfort and convenience when compared with conventional buses, while the use of an Intelligent Tracking System (ITS) helped increase their predictability and reliability. These are a few of the strategies to transform the negative notion of public transport in the existing regime. Summing up, the BRT experiment in Ahmedabad adopted both 'fit-and-conform' as well as 'stretch-and-transform' strategies of empowerment.

Bus Rapid Transit in Bangkok

If cities are characterized by the modes of transport that dominate everyday mobility, then Bangkok can be viewed as a traffic-saturated 'bus city' (Barter 1999) where most people depend on inadequate public buses or informal bus services (minivans, converted pick-up trucks, and so on). While the privileged middle-class car drivers are the root cause of increasing congestion, buses still constitute an important and affordable option for many Bangkokians.[9] In the 1980s, Bangkok had a reasonable network of bus-only lanes but with rapidly growing car ownership, the network became increasingly ineffective due to a lack of enforcement and faded into oblivion (Marler 1982; Tanaboriboon 1992). Thirty years later, faced with massive congestion and an increasingly pervasive regime of unrestrained (car-based) motorization, a number of actors set themselves the task of rebuilding a niche for priority bus-based transport. Under the auspicious label of BRT, they have recently created a system based on fancy air-conditioned buses, smart card payment, elevated boarding platforms and, of course, dedicated busways. The story of implementing the pilot route of the Bangkok BRT system in 2010 can be seen as struggle to 'stretch-and-transform' the regime dimension of technology and infrastructure.

Around the turn of the millennium, inspired by a successful and revolutionary BRT system in the city of Bogota, Colombia, the idea of a new comprehensive bus system with bus-only lanes became popular among transport specialists and policy makers in Bangkok. Two different coalitions of experts developed their own BRT plan. One group was mandated by a national-level authority (OTP), the other group by a city-level authority (BMA). Because both authorities controlled different parts of the city's road infrastructure, it was unclear who would be in charge of designing and implementing the new system. Eventually the city-level authority won out and their ideas were embraced by Apirak Kosayodhin, a politician who used the promise of a BRT system as part of his electoral platform to successfully run for governor of Bangkok.

When the first pilot route was opened in 2010, it was difficult to lay claim to the allocated road space in practice. Some motorists in Bangkok were impatient with the heavy traffic and would use the BRT-only lane. The police often turn a blind eye to this kind of behaviour, because of their everyday battle with chronic congestion, much the same as they had done with the busways of the 1980s. The city's heavy traffic burden and constant gridlock created a situation in which police officers spent much of their time trying to relieve traffic jams. They viewed it as their mandate to 'make the traffic flow' and to them the idle BRT-only lane offered additional space. Owing to objections by car drivers and police, certain stretches of BRT-only were soon no longer formally off-limits to general traffic during rush hour (Figure 9.4). 'Bangkok BRT never reached its full potential … it functions like a cripple', complained one of the interviewed consultants responsible for its design and implementation, 'the problem is that we have compromised too much on dedicated road space'.

The BRT-only lane, which was designed and constructed to ensure that buses could move quickly and unimpeded, is a defining feature that makes or breaks a BRT system (ITDP 2013). This type of dedicated infrastructure was incompatible with the paradigm of unrestrained motorization and the idea that a single-occupant car has the same right to the road as a bus filled with passengers. In the struggle to 'stretch-and-transform' technologies and infrastructures, BRT represented an alternative socio-technical configuration that conflicted with the aims of car drivers (whose road space was taken away) and the traffic police (whose mindset was

Figure 9.4 Bangkok BRT bus stuck in traffic in its own lane.

geared to regulate mixed traffic situations). In a situation where old routines of regulating traffic proved obdurate and where a growing number of middle-class car drivers wielded considerable power, the struggle for road space and a transition to infrastructural systems based on a different logic provided a significant challenge.

Conclusion

In this chapter, we have summarized four distinct mobility experiments in three Asian cities, with an aim to unpack the ways in which niche actors try to navigate tensions emerging from a mismatch with regime structures. Experiments compete against complex, multi-dimensional regimes of urban space, suggesting that navigational strategies occur on various dimensions. This chapter has demonstrated that navigational strategies are different across regime dimensions – that is, whilst experimental actors may follow a fit-and-conform strategy in one regime dimension, they may follow a stretch-and-transform dimension in another, depending on the particularities in each case and how these are assessed by the actors involved in experimentation. Nevertheless, a pattern is emerging from the analysis. Stretch-and-transform strategies occurred in the technology and infrastructure dimension (BRT experiments in both cities) and cultural significance dimension (e-rickshaws and BRT in Ahmedabad). In the case of public policy and political power, the strategies found are fit-and-conform (e-rickshaws and motorcycle taxi-meters) in nature. An explanation could be that niche actors perceive stretching-and-transforming political power to be too risky, as suggested by both cases in Bangkok. Whether or not a socially just sustainability transition in India and Thailand is possible without rebalancing political power, is a question for future research.

Notes

1 Starkey (2000) defines a short distance as 0.5–5.0 km.

2 Estimates in 2002 revealed that there were about 2 million cycle rickshaws on Indian roads carrying 6 to 8 billion passenger kilometres per year (Rajvanshi 2002). More recent estimates suggest that there are about 10 million cycle rickshaws in India (SMV Wheels Pvt. Limited 2011).

3 Drivers of rickshaws are colloquially referred to as 'rickshaw-pullers'. However, in this article we use the term drivers.

4 The launch on 2 October was the birth anniversary of Mahatma Gandhi. The inauguration was hosted by the Minister of Science and Technology, Government of India and the Chief Minister of New Delhi. The launch was a highly visible affair that demonstrated the political commitment and involvement of these national institutions.

5 CMVR is the legal and regulatory framework that guides issues such as driving licences, motor vehicle registration, and traffic control in India.

6 The CMVR, administered by the national and sub-national governments, is more or less uniform across the country. However, individual municipal authorities have specific regulations that are outside the CMVR.

7 Today, fares are informally standardized or determined through haggling.

8 The process involved citing examples of cities like Curitiba, Sao Paulo, and Beijing where BRT systems have been successful.

9 Over the last 15 years, Bangkok also developed two fast, efficient urban rail systems to cater to a privileged audience and 'splinter' the infrastructural city (Graham and Marvin 2001). In terms of the number of trips, bus-based modes (36 per cent) are far more prevalent then urban rail (4 per cent) in Bangkok (IBP 2013).

References

Bai, X., Roberts, B. and Chen, J. (2010). Urban sustainability experiments in Asia: patterns and pathways. *Environmental Science and Policy*, 13: 312–325.

Banister, D. (2011). Cities, mobility and climate change. *Journal of Transport Geography*, 19: 1538–1546.

Barter, P.A. (1999). *An International Comparative Perspective on Urban Transport and Urban Form in Pacific Asia: The Challenge of Rapid Motorisation in Dense Cities.* PhD Dissertation, Murdoch University, Australia.

Begum, S., and Sen, B. (2004). Unsustainable livelihoods, health shocks and urban chronic poverty: rickshaw pullers as a case study. *Chronic Poverty Research Centre Working Paper*, (46).

Bulkeley, H.A., Castán Broto, V. and Edwards, G. (2015). *An Urban Politics of Climate Change: Experimentation and the Governing of Socio-Technical Transitions*. London: Routledge.

Chakravarty, A. (2014). *Last Mile Chaos*. [Electronic]. Available: http://www.downtoearth.org.in/content/last-mile-chaos [1 December, 2014].

CMERI (2010). *Soleckshaw (Solar-Electric Rickshaw): A CSIR-CMERI Technology*. Central Mechanical Engineering Research Institute. [Electronic]. Available: http://www.cmeri.res.in/oth/news.htm [29 July, 2015].

Delhi High Court (2014). *Judgement in matter of writ petition: Shanawaz Khan vs. New Delhi Municipal Corporation*, New Delhi: Delhi High Court.

Furlong, K. (2010). Small technologies, big change: Rethinking infrastructure through STS and geography. *Progress in Human Geography*, 35: 460–482.

Gakenheimer, R. (1999). Urban mobility in the developing world. *Transportation Research Part A: Policy and Practice*, 33: 671–689.

GEA (2012). *Global Energy Assessment – Towards a Sustainable Future*. Cambridge/Laxenburg, Austria: Cambridge University Press/The International Institute for Applied Systems and Analysis (IIASA).

Geels, F. (2002). Technological transitions as evolutionary reconfiguration processes: a multi-level analysis and a case study. *Research Policy*, 31: 1257–1274.

Geels, F.W., Kemp, R., Dudley, G. and Lyons, G. (2012). *Automobility in Transition? A Socio-Technical Analysis of Sustainable Transport*. New York: Routledge.

Ghosh, D., Roy, J., Raven, R. and Saha, S. (2013). Motorization of cycle rickshaws (vans) in India: analysis of a few experiments. *GCP-JU Working Paper: GCP-JU/WOTRO/05/01*.

Goswami, U. (2012). UN Climate Change Negotiations 2012: Ahmedabad's Bus Rapid Transit System to be showcased by United Nations. *The Economic Times*, 15 November 2012.

Graham, S. and Marvin, S. (2001). *Splintering Urbanism: Networked Infrastructures, Technological Mobilities and the Urban Condition*. London: Routledge.

Gwilliam, K. W. (2002). *Cities on the Move: A World Bank Urban Transport Strategy Review*. Washington, DC: The World Bank.

Hildebrand, M., Kanaley, T. and Roberts, B. (2013). *Strategy Paper: Sustainable and Inclusive Urbanization in Asia Pacific*, New York: United Nations Development Programme.

Hoogma, R., Schot, J. and Truffer, B. (2002). *Experimenting for Sustainable Transport: The Approach of Strategic Niche Management*. London: Spon Press.

IBP (2013). *Thailand Transportation Policy and Regulations Handbook Volume I: Strategic and Practical Information*. Washington, DC: International Business Publications.

IPCC (2014). *Mitigation of Climate Change. Contribution of Working Group III to the Fifth Assessment Report of the Intergovernmental Panel on Climate Change*. Cambridge: Cambridge University Press.

ITDP (2013). *The BRT Standard*. New York: Institute for Transportation and Development Policy.

Karvonen, A. and van Heur, B. (2014). Urban laboratories: experiments in reworking cities. *International Journal of Urban and Regional Research*, 38: 379–392.

Khan, J., Hassan, T. and Shamsad, S. (2012). Socio-economic profile of cycle rickshaw pullers: a case study. *European Scientific Journal*, 8: 310–330.

Kumar, A. (2011). *Understanding the Emerging Role of Motorcycles in African cities: A Political Economy Perspective*. Sub-Saharan Africa Transport Policy Program (SSATP) Discussion Paper.

McMichael, A.J. (2000). The urban environment and health in a world of increasing globalization: issues for developing countries. *Bulleting of the World Health Organization*, 78: 1117–1126.

Mahadevia, D., Joshi, R. and Datey, A. (2013). *Low-Carbon Mobility in India and the Challenges of Social Inclusion: Bus Rapid Transit (BRT) Case Studies in India*, Denmark: UNEP Risø Centre on Energy, Climate and Sustainable Development.

Maji, P. K., Banerjee, P. S., Banerjee, A. J. and Maity, S. (2010). Electric motor-assisted pedal driven tricycle. *International Journal of Electric and Hybrid Vehicles*, 2: 202–210.

Markard, J., Raven, R.P.J.M. and Truffer, B. (2012). Sustainability transitions: an emerging field of research and its prospects. *Research Policy*, 41: 955–967.

Marler, N.W. (1982). *The Performance of High-flow Bus Lanes in Bangkok*, Crowthorne, UK: Transport and Road Research Laboratory.

Moriarty, P. and Honnery, D. (2008). Low-mobility: the future of transport. *Futures*, 40: 865–872.

Rajvanshi, A. (2002). Electric and improved cycle rickshaw as a sustainable transport system for India. *Current Science*, 83: 703–707.

Sengers, F. and Raven, R. (2014). Metering motorbike mobility: informal transport in transition? *Technology Analysis and Strategic Management*, 6: 453–468.

Sengers, F. and Raven, R. (2015). Toward a spatial perspective on niche development: the case of Bus Rapid Transit. *Environmental Innovation and Societal Transitions*, 17: 166–182.

Sheller, M. (2012). The emergence of new cultures of mobility: stability, openings and prospects. In Geels, F., Kemp, R., Dudley, G. and Lyons, G. (eds), *Automobility in Transition? A Socio-technical Analysis of Sustainable Transport*. New York: Routledge, 180–202.

Smith, A. and Raven, R. (2012). What is protective space? Reconsidering niches in transitions to sustainability. *Research Policy*, 41: 1025–1036.

SMV Wheels Pvt. Limited (2011). *Dasra Social Impact: Converting Poor to Power*. [Electronic]. Available: http://www.dasra.org/n/forwebsite/dasra/Reports/SMV%20 Wheels.pdf [28 November 2014].

Starkey, P. (2000). *Local Transport Solutions: People, Paradoxes and Progress – Lessons Arising from the Spread of Intermediate Means of Transport*, Washington, DC: Rural Travel and Transport Program (RTTP), The World Bank.

Tanaboriboon, Y. (1992). An overview and future direction of transport demand management in Asian metropolises. *Regional Development Dialogue*, 13: 46–70.

TIO (2014). *Experts welcome e-rickshaw guidelines*, New Delhi: *The Times of India*, 11 August 2014.

Tyfield, D. (2014). Putting the power in 'Socio-Technical Regimes' – e-mobility transition in China as political process. *Mobilities*, 9: 585–603.

United Nations (2014). *World Urbanization Prospects: The 2014 Revision*, New York: United Nations, Department of Economic and Social Affairs, Population Division.

Valderrama, A. and Vogel, N. (2014). Transitioning to a low carbon society? The case of personal transportation and urban form in Copenhagen: 1947 to the present. *Transfers*, 4: 4–22.

Verbong, G., Geels, F. and Raven, R. (2008). Multi-niche analysis of dynamics and policies in Dutch renewable energy innovation journeys (1970–2006): hype-cycles, closed networks and technology-focused learning. *Technology Analysis and Strategic Management*, 20: 555–573.

10 Urban science networks and local economy

The case of Newcastle upon Tyne

Gareth Powells and Lynsay Blake

Introduction

In this chapter, we examine the role of city science projects in post-industrial urban economies and the drivers of their co-production. We argue that urban experiments function as a result of their ability to weave shared visions of urban futures into which multiple groups and organisations can invest. In developing this argument, we draw upon research undertaken within and around Newcastle upon Tyne in the United Kingdom. Newcastle is England's most northerly city, positioned on the north bank of the River Tyne and nine miles west from the North Sea coast. The city is in the metropolitan county of Tyne and Wear in the north-east of England. While the city itself has a population of 280,200, Tyne and Wear has a combined population of 1,104,800 (Office for National Statistics 2011).

The research undertaken to inform this chapter includes both desk-based studies of literature and policy documents as well as participatory and qualitative research with a group of actors in the city involved in a number of urban experiments focused on sustainability. We refer to this group as the Newcastle Urban Sustainability Science Network (NUSS Network). We first discuss the way that contemporary research characterises urban science–policy interfaces, and the ways in which experiments feature in these relationships. We then draw on the research materials to develop an account of Newcastle's science network to examine the drivers for urban experiments.

Urban science and sustainable urban development

Newcastle is a famously compact city in which a large indoor shopping mall and premier league football stadium located in the city centre form part of a (in) famous cultural economy of retail, sport and partying (Vall 2001; Chatterton and Hollands 2002; Gratton and Henry 2002; Shaw 2015). However, the city also has a history of creating space for science in the heart of the urban landscape and in the local economy. In 1963, an education precinct was formed in which the newly independent Newcastle University was allocated space alongside what was the Polytechnic College (now the University of Northumbria) on the realisation that there were economic benefits associated with having a thriving university placed within the city. This was thought to be both in terms of the 'food and books' that would be sold but also in terms of promoting the city's position in knowledge and innovation economies. The City Council leader of the time, T. Dan Smith, noted that an urban university would be a 'tremendous source of industrial attraction to the city and region' (quoted in Goddard 2014: 3). The city continues to make new spaces for science not only because of the value attributed to the knowledge outputs of co-produced urban science but also because the relationship between 'town and gown' is recognised as an important dynamic in the urban economy.

Authors like Whitehead (2015) and Marvin and Hodson (2014) have explored the political economy of sustainable urban development, arguing that opportunities as well as tensions 'derive from the role of cities as hubs of economic development and environmental management' (Whitehead 2015: 196). However, the role of universities is explored infrequently. As sustainable urban development becomes more a matter of local, place-based experimentation than the planned roll-out of national plans (Bulkeley 2013), the connections between sustainability, science spaces and the urban economy are beginning to attract the attention of scholars building on early work on the economic geography of technopoles and science-parks (Benko 2000) as important features of regional innovation systems. More recently, Coenen and others have made clear that universities are increasingly being expected to participate in more 'mode-2' (problem-based) knowledge creation. As a result, they are being encouraged to be more entrepreneurial, to 'contribute directly to a more dynamic development of the business sector – in addition to the traditional roles of teaching and research' (Coenen 2007: 806).

This change in the position of the university in the city and the insertion of scientific methods into the practices of urban governance have now been nationally institutionalised in the UK. The national government created a new ministerial post in July 2015: Minister for Universities, Science and Cities, with Greg Clarke MP, formerly Minister for Cities, being the first individual appointed to this position. By merging the Universities and Cities portfolios into a single post, the UK government has explicitly placed local economic growth, smart cities, research and science into a single, integrated set of national responsibilities. This reflects work already underway to integrate these areas of policy in several British cities such as Nottingham[1] and Birmingham,[2] both of which have mature science-city spaces and networks that figure prominently in their universities' brands and their cities' economic narratives. In developing our analysis, we extend Coenen's argument to suggest that urban science spaces are one of the principal ways in which universities are entering into urban economic management and that collaborative environmental sustainability science is a primary focus of such projects. Urban science spaces are a means of directly inserting cities into global innovation and science networks, whilst also creating visible, globally connected places in the urban landscape for the co-performance of science, environmental governance and service provision. In Newcastle's case, rather than an out-of-town innovation park or technopole, a large central site in the city has been dedicated to 'urban sciences', contributing to the centrality of Newcastle's post-industrial economy.

In our case study of Newcastle upon Tyne, we find that urban science is not a bilateral endeavour restricted to the municipal authority and the university but is instead co-performed by a network of urban actors who together constitute an urban science network. Using the concept of networks to characterise the policy process as one populated by actors from different organisations working in loosely and self-organised ways to bring about changes in policy has been a common feature of work on environmental sustainability over recent decades (Chilvers and Evans 2009). This has led to the development of a rich body of research that theorises the relationships between science and policy, with writers like Marsh and Rhodes (1992) and Hajer and Wagenaar (2003) describing the reconfiguration of the roles, responsibilities and capacities of state institutions that has resulted in a smaller role for the state and a larger role for non-state actors in governing processes. This can be summarised as a shift from government (by state institutions) to governance, and represents a widening of participation in governing processes (Rhodes 2007) through which universities have been recognised to have important roles to play in policy processes. In the following sections, we develop an analysis of the spaces and forms that this involvement takes. Although these writers among others, have established that the university has entered into the political sphere and the policy process, the reverse has been shown to be equally evident: that scientific processes are political and that any delineation between these domains is both artificially produced and always in need of re-establishment (Jasanoff 1987). The dynamics between science and urban political processes at the city scale have led to the emergence of collaborative, co-produced mode-2 science as a basis for newly reconfigured urban science–policy relationships; one that in many ways seeks to enable new forms of policy-making based on experimental processes rather than hard evidence.

We use the definition of 'experiments' as outlined by Trencher and colleagues (2014), who describe them as planned initiatives that take a combined social and technical approach. Here, it is assumed that no individual participant within the network possesses the knowledge, resources, or ability to single-handedly bring about socio-technical change. Therefore, the strength of the network is in the potential for mobilisation of 'knowledge, capabilities and resources between partners' (Trencher *et al.* 2014: 154). In such experiments, the network provides actors with the opportunity to frame and address issues of local and regional importance but which are applicable to wider society and as a result are able to attract both international interest and connect actors to global knowledge economies.

Newcastle, sustainability and the urban sustainability science network

Newcastle upon Tyne has a rich history of innovation relating to science, engineering and energy systems in particular, with historic figures such as Joseph Swan (inventor of electric lighting), William Armstrong (engineer, industrialist and the first scientist to join the House of Lords) and George Stephenson (inventor of the steam locomotive and first inter-city rail line) featuring heavily in the narrative of the city. Interesting to note, however, is that each of these figures is remembered for both his scientific and his industrial accomplishments; each brought academic prestige, as well as economic rewards to the city and wider region. The actions of these innovators, amongst others within the north of England, gave the region its historic reputation as 'The Northern Powerhouse' in the nineteenth and early twentieth centuries. This definition has re-emerged in twenty-first century politics in the form of a regionally interconnected 'Northern Powerhouse', with science, innovation and cities figuring centrally in contemporary discourses about the future of post-industrial northern economies in which city-level state institutions are positioned to gain more devolved powers from Westminster (Osbourne 2014).

Despite the fact that in recent years the energy systems innovation landscape within Newcastle has undergone significant change and consolidation, the city continues to project an image of itself as a place which attracts large international applied energy science investors who have developed and expanded their interests within the north of England, as well as a myriad of spin-outs, SMEs and micro-businesses (Adonis *et al.* 2015). This was recognised and formalised in 2004 when Newcastle was identified as one of six UK 'Science Cities', which led to the inception of the Newcastle Science City Partnership, a special purpose vehicle to provide cross-city support between private and public sectors, supporting investment in the city's key science sites including the Centre for Life (a life sciences lab and public engagement centre located adjacent to the city's central train station) and most recently Science Central, a new urban quarter that integrates the university into the general fabric of the city. This is in contrast to most science parks that are located on the urban periphery of host cities.

In this context, the city has embraced local, co-created science to work with urban partners to produce outputs of mutual value, including both delivery and measurement of impact of its action plan. This has been far from a smooth, linear process but, over a period of several years, a group of actors have constituted a flexible and evolving network that has initiated a number of experiments. Members of the Newcastle Urban Sustainability Science Network (NUSS Network) are presented in Figure 10.1. They include local actors such as Newcastle City Council, Newcastle University, Northern Gas Networks, Northern Power Grid, Your Homes Newcastle, Northumbrian Water, Newcastle Science City and National Energy Action, as well as more remote partners positioned in national and international networks, such as large multinationals (e.g. Siemens) and SMEs (e.g. Agility Eco). In the following sections, we argue that the participants in the NUSS Network see membership as being both self-serving and driven by a desire for sustainable urban change.[3] The relationships and paths of exchange in the NUSS Network are varied, based on formal and informal arrangements, and are dependent on the resources and capabilities of the partner organisations involved and the identified needs of these organisations.

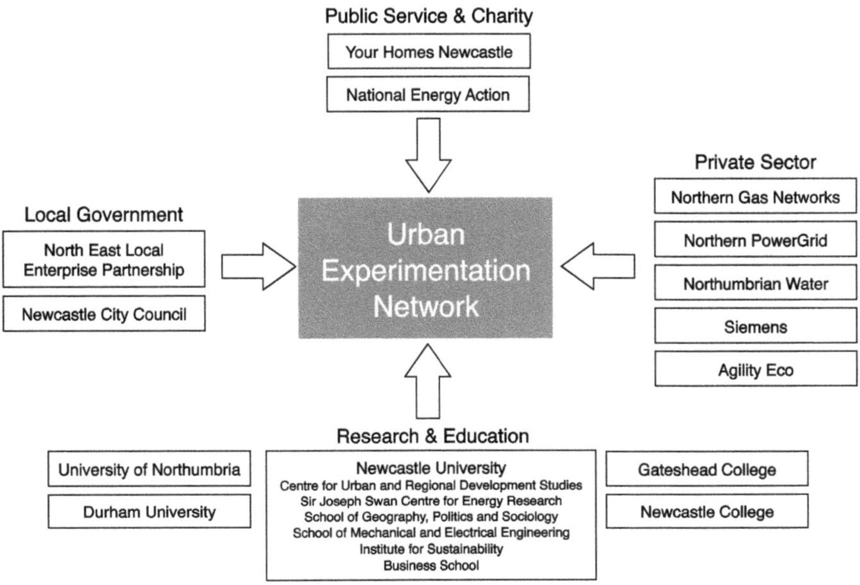

Figure 10.1 Members of the Newcastle Urban Sustainability Science Network.

Three NUSS Network projects

We now introduce three key projects of co-production of the NUSS Network. Each project is distinctive in many ways and they resist categorisation as 'types' of experiments, projects, initiatives or otherwise. However, their common link is the central role played by members of the NUSS Network in collectively pursuing opportunities as a group of urban partners.

Project 1: Science Central

Newcastle's most high profile and distinctive experimentation initiative is Science Central, a major 24-acre urban development situated on a brownfield site in the centre of Newcastle. The shared vision for this development is to provide a demarcated, mixed-use space in the city – a new kind of learning precinct, to use T. Dan Smith's language – in which the public sector, academia, industry and communities can co-locate, live and work. This is seen as a flagship asset to host a range of possible urban science experiments with an emphasis on sustainability, innovation and social renewal, guided by the core objectives of the university (to be a civic university) and the city (to be a working city). Science Central is explicitly aimed to attract inward investment to the city as can be seen in the investor events accompanying each phase of its launch and in its stated goal to 'support a thriving community, rewarding jobs and ground breaking scientific advances' (Newcastle Science City Partnership 2015).

The first building on site, funded through the European Union's Regional Development Fund and Newcastle City Council, provides contemporary office space for firms working on a range of future city challenges including energy, transport, big data and cloud computing. This space is designed to be an iconic building for urban sustainability in Newcastle, featuring a living wall, bee hotels, planted sedum roofs, rainwater harvesting and a series of adjoining public squares.

The second building to be established on site, the Urban Sciences Building (USB), is designed as a multi-purpose facility that includes laboratory functions such as the Urban Observatory and Decision Theatre. The Urban Observatory will collect diverse datasets from across the city that can then be integrated across varying spatial and temporal scales. This real-time data will allow the city to be analysed and explored in order to improve the understanding of interactions between the city's energy, water, transport, waste and digital control systems, thereby providing new insights into how cities operate and behave as dynamic highly complex, socio-technical systems. The Decision Theatre will provide a space where urban systems (including but not limited to data from the Urban Observatory) can be analysed by stakeholders including utility companies, emergency services, local authorities, the general public and government bodies. The aim is to facilitate public engagement around urban sustainability and encourage multi-stakeholder groups to explore and debate in an immersive, data-rich space.

Project 2: Energy Technologies Institute – Smart Systems and Heat Demonstrators (ETI-SSH)

In this project, the NUSS Network is part of an urban experiment in which the ETI-SSH programme (to be incorporated within the new UK Energy Systems Catapult) will work across three competitively selected demonstrator local authorities (Newcastle, Manchester and Bridgend) to explore the most economically, socially and technically viable options for heating homes in the UK in order to realise policy aims connected with decarbonising domestic energy use by 2050 (Ekins *et al.* 2013). Combined, these local authority regions will represent the range of UK-based housing archetypes, with their communities acting as 'labs' to represent the UK's residential communities.

Initially, data from these local authorities is being incorporated into the development, testing and validation of the Energy Path Network tool. This tool is designed to inform the most cost-effective energy transition plan for a specific area, taking into account multiple factors including building stock, heat demand and existing utility infrastructures. Local demonstrator projects will then take part in the experimental process to examine the social, technical and economic efficiency and efficacy of a number of retrofit technologies across a range of housing types drawing on the city's housing stock. This project will use data collected from interventions conducted 'out there' in the urban environment using buildings that represent the UK's wider urban population. Facilities such as Science Central's Decision Theatre in the USB will be used in collaborative analysis activities.

Project 3: Sustainability Multi-Storey Communities (SMS)

The members of the NUSS Network have initiated a project focusing on the challenges, complexity, and multi-stakeholder interdependence of multi-storey communities and how previous approaches to addressing social and environmental sustainability in high-rise buildings have been constrained. This project brings together utility providers, local authorities, community organisations and housing providers to identify positive sustainability opportunities for multi-storey communities that could arise from differently configured social, technical, regulatory and commercial relationships between the existing infrastructures. The project uses data-capture instruments to record flows into and out of five high-rise tower blocks in the city and uses ongoing community engagement activities as part of a participatory qualitative research process running in parallel with the quantitative data capture. As well as layering lab-like characteristics onto this already existing and spatially bounded community, the project is explicitly inviting partners into new spaces for co-production in the university. The Business School's Living Lab facility exists in parallel with the USB to facilitate detailed explorations of interlinked problems of water, waste, energy, communications and housing service innovation. The Living Lab is an additional space for data-enriched deliberative decision-making housed at the Business School rather that in Science Central.

Taken together, emerging modes of experimentation in the NUSS Network are creating modes of experimentation and analysis in at least three spatially distinct ways. First, there are already existing, bounded spaces which are being newly instrumented to study particular aspects of urban life (such as clusters of high-rise buildings). These are places in the city that are being *labbed*. Second, there are web-like processes in which the city's spatially extensive existing and emerging sensor networks (traffic, energy use, air quality, rainfall, and so on) are being tapped-into and their data sent to nodes like the Urban Observatory in the USB building. Here, the city's existing sensory instruments are being *webbed*. Last, there are spaces of engagement, discussion and experimental play, in which members of the network can meet to interact with data and visualisations through simulations scenarios, forecasts, and virtual interventions in spaces like the Living Lab and the Decision Theatre, which are meeting rooms augmented by data, dashboards and visualisations. Here actors in the city are being *engaged*.

Having set out these processes through which elements of the city are *labbed*, *webbed* and *engaged*, we now turn to qualitative data created in a participatory project with members of the NUSS Network to examine in more detail the drivers for involvement in these initiatives.

Drivers of urban science: analysis of stakeholder data

Representatives of major groups of actors within the NUSS Network were asked about their motivations and drivers for involvement in the network and the extent to which the aforementioned projects have the potential to lead to the realisation of radical social and environmental change. Common motivations and drivers for involvement in urban experiments were shared across the network and related to achieving organisational objectives, increasing access to current and future funding streams, as well as research opportunities.

Research opportunity

Academic partners reported a strong link to organisational objectives that also matched their own personal commitments to a style of scientific practice. They felt that membership of the network was key in addressing Newcastle University's actions in being a 'civic university' (one of the key aspects of the institution's projected identity) and that this was also key to, 'providing a route to changing the environment and lives of the people in Newcastle city'. This aligned closely with the perspective of public sector partners who felt that the network allowed exploration of societal issues from multiple viewpoints. One member of the network explained that involvement in the NUSS Network was important to him because it meant that 'complex problems can be understood from different aspects and the proposed solutions are not only applicable technically, but also remain relevant and sympathetic to the existing history and heritage of the place'.

Organisational and commercial objectives

Urban experiments were also viewed by respondents as a means of meeting core organisational objectives and these are often closely linked with the development of future strategy. These statements were clearly expressed as forward-looking, aspirational stances that were shaped by the nature of the organisation and by the commitments it has made or the responsibilities it has to comply with obligations made by relevant statutory bodies (such as carbon emission reduction targets). For example, a representative of Newcastle City Council stated that these networks represent a unique opportunity to 'serve the key priorities of the city namely being a working city, providing decent neighbourhoods, tackling inequalities, and being a fit-for-purpose council', whilst simultaneously addressing wider interconnected challenges/wicked problems of 'climate change, fuel poverty, economic development and growth'. In similar alignment with organisational goals, a representative of a social-housing provider stated that the purpose of engaging in this process was to 'maximise the quality of service delivered to our customers' and to formulate future 'best practice' to position themselves in a leadership position within the UK social-housing sphere. This was similar to the views of utilities companies and commercial partners who stated that the project allows them to 'develop a better understanding of expectations around energy usage now and into the future to help inform future strategy and plans'. They went on to say that participation in urban experiments was initiated and sustained by 'largely commercial' drivers. Non-monopoly firms (those not holding licensed regional monopolies for operation of network infrastructure) added that projects in the NUSS Network helped establish their position as 'thought leaders in areas like this and differentiates us from other players in the market' and helps them 'meet other potential clients'.

The NUSS Network was thought to enable access to funding and finance in several ways. This was often expressed in terms of utilising the network partnerships to collaboratively access national and international funding streams. One way in which this worked was to enable members to access research funding available exclusively to collaborative networks or as a way to create an indirect link to a funder that would otherwise be inaccessible to some members of the network. A second way in which the network was reported to be financially advantageous was by making it possible to coordinate or match funding streams by co-investing in an initiative to realise synergistic but separate returns (by co-funding a retrofit, for example). Additionally, third-sector organisations stated that these relationships allowed 'diversification of income streams' by matching the 'skillset of our own research team' with those of partners in order to participate in projects that would otherwise be outside their scope. This was articulated further by Newcastle City Council who made it explicitly clear that the post-2008 era of austerity has made partnerships of this kind even more financially important; 'collaboration is key for Newcastle City Council as internal resources are becoming restricted due to significant budgetary cuts and austerity' and therefore they see it as increasingly important to work with key network partners 'to attract investment and revenue funding into the city'.

This confirms that in the post-industrial, post-austerity context, for Newcastle to be what it wants to be – a working city – collaborative investment in science and innovation is seen as important not only because of the imagined sustainability benefits to be derived from the content of the experiments but also because of the economic rewards thought to come from improving Newcastle's position in global science economies.

Urban optimisation

The respondents also noted that the NUSS Network provides the city's key actors with an opportunity to establish a core set of aims and objectives for sustainable urban development (albeit ones with experimentation and trial-and-error built-in) that are based on local actor needs. Without a network which persists from funding round to funding round and from national government to national government, the city would have to adopt a more reactive stance and revise its objectives in response to the priorities of other (usually national) bodies. Through the NUSS Network the city can develop a clearer sense of its own priorities that can adapt to rather than be continually re-born in response to each opportunity. Furthermore, the continuity and cross-sector nature of the network was felt to help guard against the threat of sub-optimal outcomes. As one respondent states:

> If experiments and learning are carried out on a project-by-project basis, there is a risk that the overall urban system becomes sub-optimal as it is the aggregation of a number of locally optimal projects where interactions have not been fully considered.

Some network members felt that the network itself helped guard against this, others acknowledged that sub-optimality is difficult to rule out. A more critical interpretation suggests that both of these perspectives share a fundamental belief in the possibility of an optimal solution that can overcome all sub-optimalities through better, bigger, more multi-dimensional data and new forms of analysis. This is evident only in the sub-text of the research we have conducted but, nonetheless, there was a definite shared belief that co-produced urban science would yield solutions that have been thus far inaccessible through conventional working practices. As a respondent argued:

> Delivering sustainable, real and radical social and environmental change is a very difficult thing to achieve. However, only by working in partnership with a broad range of forward-thinking stakeholders can the complexity of an issue(s) be understood and therefore appropriate solutions be identified and implemented.

However, the respondents recognised that urban science and experimentation is challenging. There was general recognition that these projects are intellectually stimulating. A respondent noted, 'close examination of these communities could provide lessons that could be transferred to similar projects and more importantly scaled up to district or city level'. However, some partners expressed uncertainty about the issue of inference and extrapolation form the urban 'case' to a wider population. Another respondent asked:

> How can you make a project that solves a specific problem in the short term in Newcastle lead to an impact case study, top quartile journal papers and international reputation? This is something that all academics who do applied research grapple with.

Discussion

In the NUSS Network, and Science Central in particular, we can see the politics of urban science in Newcastle being re-awakened through the re-establishment of the 'town-and-gown' partnership at the heart of attempts to govern the city's sustainable urban development. However, this is not a bilateral 'special relationship' but rather a network that pulls in a range of actors involved in the governance of the city's post-industrial economy and its infrastructure founded on the shared assumption that to create spaces for science is to create spaces for high value employment, inward investment and income opportunities as well as 'learning'. The imagined future of the city as a prosperous science 'centre' is encapsulated in a statement of Science Central's vision and purpose which confirms that the impetus to make Newcastle a science city once again is not simply a top-down designation but a locally driven commitment:

> Combining cutting-edge architecture with new public spaces, world-renowned scientific expertise and leading-edge companies, it will be an innovation hub where investors, businesses, entrepreneurs, students, scientists and citizens collaborate to plan and develop solutions for tomorrow's cities. It will be used as a living laboratory where solutions can be tested, demonstrated and commercialised, creating a lasting legacy of science and innovation for the North East.
>
> (Newcastle Science City Partnership 2015)

This potent cocktail of iconic architecture, low rent (by national standards), and urban identity formation, underwritten by an appeal to 'science' and the laboratory, make for a compelling case; one which is being made to mobile capital and businesses as well as scientists and other members of the creative class (Florida 2005). Through Science Central and other projects initiated by the NUSS Network, the city challenges current readings of experimental governance. Far from being only a feature of environmental governance as it is most often understood (Hoffmann 2011), the case of Newcastle indicates that experimental governance is also a practice at the heart of post-crisis urban development and economic management. By braiding environmental management commitments together with urban development imperatives, the cultural capital of science, the laboratory and the scientific method, city networks such as the NUSS Network are able to form new powerful coalitions capable of realising ambitious imagined urban spaces. While we have highlighted Newcastle's history of compressing its cultural and knowledge economies into a compact city centre, locating Science Central alongside the football stadium, shopping mall, museums, pubs, hotels and Newcastle University's Business School confirms that the various spatialities of urban labs in Newcastle are intended to be an integrated part of the post-industrial urban economy rather than an array of observation towers from which to study it.

Underpinning the work of the NUSS Network, however, is the assumption that complex sustainability issues can be addressed at local to regional levels by collaboration and experimentation and that this can potentially be replicated at a wider scale. While the NUSS Network, among others, is attempting to achieve this through an inclusive process involving co-design and co-production, its ultimate consequences are uncertain. At present, there have been few studies to critically evaluate the effectiveness of urban experiments, science spaces and living labs in bringing about the kinds of radical change required to address the scale and breadth of sustainability challenges of our times (Mauser *et al.* 2013). Whether the best practice of participatory urban science will be good enough remains to be seen. What is clear however is that the city is being positioned by academics and the state as both the main subject and object of experimentation as well as a major engine of economic performance and governance.

Notes

1 Nottingham's science-city network can be found at http://science-city.co.uk/.
2 Birmingham's science-city network can be found at http://www.birminghamsciencecity.co.uk/.
3 This chapter focuses on energy related aspects of the NUSS network but the reach of the network is wider and involves projects which span social and technical perspectives on the interconnected sustainability challenges of integrated infrastructure systems, resource production, consumption and waste (including water, transport, energy, food and earth systems engineering).

References

Adonis, L.A., Mottram, H., Curry, L.D., Hutton, W., Rosewell, B. and Ruffer, J. (2015). *The North East LEP Independent Economic Review*. [Online]. Available: http://nelep.co.uk/wp-content/uploads/2015/02/NE-Economic-Review-Evidence-Base-Summary.pdf [5 May 2015].

Benko, G. (2000). Technopoles, high-tech industries and regional development: a critical review. *GeoJournal*, 51: 157–167.

Bulkeley, H. (2013). *Cities and Climate Change*. London: Routledge.

Chatterton, P. and Hollands, R. (2002). Theorising urban playscapes: producing, regulating and consuming youthful nightlife city spaces. *Urban Studies*, 39: 95–116.

Chilvers, J. and Evans, J. (2009). Understanding networks at the science–policy interface. *Geoforum*, 40: 355–362.

Coenen, L. (2007). The role of universities in the regional innovation systems of the North East of England and Scania, Sweden: providing missing links? *Environment and Planning C: Government and Policy*, 25: 803–821.

Ekins, P., Keppo, I., Skea, J., Strachan, N., Usher, W. and Anandarajah, G. (2013). *The UK Energy System in 2050: Comparing Low-Carbon, Resilient Scenarios*. London: UK Energy Research Centre.

Florida, R.L. (2005). *The Flight of the Creative Class: New Global Competition for Talent*. New York: Harper Business.

Goddard, J. (2014). *Newcastle as a City of Learning*, 9 January 2014. [Online]. Available: http://www.ncl.ac.uk/curds/news/item/professor-john-goddard-giveslecture-in-newcastle-lit-phil-series [12 February 2015].

Gratton, C. and Henry, I. (2002). *Sport in the City: The Role of Sport in Economic and Social Regeneration*. London: Taylor & Francis.

Hajer, M. A. and Wagenaar, H. (2003). *Deliberative Policy Analysis: Understanding Governance in the Network Society*. Cambridge: Cambridge University Press.

Hodson, M. and Marvin, S. (eds) (2014). *After Sustainable Cities?* Routledge: London.

Hoffmann, M.J. (2011). *Climate Governance at the Crossroads: Experimenting with a Global Response after Kyoto*. New York: Oxford University Press.

Jasanoff, S.S. (1987). Contested boundaries in policy-relevant science. *Social Studies of Science*, 17: 195–230.

Marsh, D. and Rhodes, R.A.W. (1992). *Policy Networks in British Government*. Oxford: Oxford University Press.

Mauser, W., Klepper, G., Rice, M., Schmalzbauer, B.S., Hackmann, H., Leemans, R. and Moore, H. (2013). Transdisciplinary global change research: the co-creation of knowledge for sustainability. *Current Opinion in Environmental Sustainability*, 5: 420–431.

Newcastle Science City Partnership. (2015). *About Science Central*. [Online]. Available: http://www.newcastlesciencecentral.com/about/ [9 July 2015].

Office for National Statistics (2011). *Census*. [Online]. Available: http://www.ons.gov.uk/ons/guide-method/census/2011/index.html?utm_source=twitterfeed&utm_medium=twitter [12 April 2015].

Osbourne, G. (2014). *We Need a Northern Powerhouse*. www.gov.uk, 23 June 2014. [Online]. Available: https://www.gov.uk/government/speeches/chancellor-we-need-a-northern-powerhouse [1 April 2015].

Rhodes, R.A.W. (2007). Understanding governance: ten years on. *Organization Studies*, 28: 1243–1264.

Shaw, R. (2015). 'Alive after five': constructing the neoliberal night in Newcastle upon Tyne. *Urban Studies*, 52: 606–616.

Trencher, G., Bai, X., Evans, J., McCormick, K. and Yarime, M. (2014). University partnerships for co-designing and co-producing urban sustainability. *Global Environmental Change*, 28: 153–165.

Vall, N. (2001). The emergence of the post-industrial economy in Newcastle 1914–2000. In Colls, R. and Lancaster, B. (eds), *Newcastle upon Tyne: a modern history*. Chichester. Phillimore and Co. Ltd.

Whitehead, M. (2015). Urban economic development and environmental sustainability. In Paddison, R. and Hutton, T. (eds), *Cities and Economic Change: Restructuring and Dislocation in the Global Metropolis*, London: Sage.

11 Grassroots experimentation

Alternative learning and innovation in the Prinzessinnengarten, Berlin

Jana Wendler

Introduction

Cities, it seems, have gone experimental: the term 'experiment' is ubiquitous, describing various kinds of artistic interventions, pioneer projects, model buildings or research collaborations. It also expresses a wider conceptual approach towards urban change that is visible in urban discussion and policy-making. At times, this experimental label refers to the uniqueness of the projects in question, or to their drive towards a radical reassessment of existing urban formations. In other cases, it denotes an open or iterative approach that focuses as much on the process as the outcome of the intervention. What unites these initiatives is their concern with the future of urban living combined with a commitment to direct practical action: not only proposing or envisioning an idea, but implementing it within the lived city. Such 'everyday utopias', as Cooper (2014: 2) calls them, 'work by creating the change they wish to encounter' as they 'perform regular daily life … in a radically different fashion'. Examples range from temporary camps exploring creative sustainability (ExperimentDays 2014) to long-term eco-squats and autonomous areas (Christiania 2004; Can Masdeu 2014) that consider themselves as labs and test sites for alternative urbanism.[1]

Experiments have already attracted significant attention in the urban context as strategic interventions and governance tools. Applied to issues such as climate change and social inequality, particularly in areas like infrastructure, transport and energy, they provide a distinctive governance approach amongst local governments, the private sector and civil society (Castán Broto and Bulkeley 2013). They are seen as both locally useful and more widely visible, responding to particular real-world problems but also acting as signposts or blueprints through which innovative policy ideas may be shared (Evans and Karvonen 2011). These discussions tend to centre on experiments that are *deliberately designed*, with a distinct mode of knowledge co-production and applied innovation focused on a specific problem.

Yet alongside an emerging view of experiments as more open-ended, heterogeneous, socio-material assemblages of learning, this instrumental perspective appears somewhat restrictive. There is a diversity in urban experiments that is not captured here. This includes experiments that *emerge organically* from existing interventions and projects: experiments that are not designed but which acquire their experimental qualities over time, and which serve as more open sources of urban learning. Many of them also foster distinctly critical angles that challenge mainstream urbanisation and propose more radical visions of the future. Paying attention to these critical, emergent, or what I propose to call 'alternative experiments' not only expands our conceptual understanding of the term but also underlines different ways in which experimentation may foster urban change.

This chapter makes a case for regarding emergent and critical alternative experiments as distinctly valuable urban interventions. To do so, it charts the experimental dynamics and wider urban role of one project, the Prinzessinnengarten in Berlin. As well as being an active community garden, the Prinzessinnengarten has earned a wider reputation as an alternative, unusual and experimental space that speaks to a range of urban issues. By exploring a number of its daily practices and ongoing projects, I will show the formation of the garden's experimentality and its resulting relevance as both an alternative space of learning and an emergent material intervention in urban policy discussion. This example highlights the intriguing and fruitful intersection of urban alternatives and experimentation, which opens up more diverse pathways of experimental learning and innovation. Following these pathways allows for a much greater diversity of ideas and approaches to become visible in urban discourses, going beyond managerial conceptions of urban change towards self-organised and more radical possibilities.

Experiments and the need for urban learning

How cities learn is a fundamental question for debates around urban change, one that is explored in the work on policy mobility and circulation (see McCann and Ward 2011; Peck 2011) and urban learning assemblages (McFarlane 2011). These studies highlight the diverse elements involved in urban policy-making, tracing in particular the emergence and adaptation of new ideas. A key dimension here are the 'mobilities-and-mutations' (Peck 2011: 1) of policy-making: urban learning is not restricted to fixed pathways of transfer and diffusion, but involves complex interactions of movements and adaptation that draw on both local dynamics and wider networks. McFarlane (2011) identifies three aspects here. One is *translation*, which refers to the relations and distributions through which learning is produced. While there is a focus on displacement and change, or how knowledge moves, translation describes this movement as always contingent and locally produced. This requires a second dimension, the *coordination* of the different spaces and actors involved in this process. Rather than following existing channels, urban learning means finding ways of dealing with the complexity and uncertainty of the urban realm through a 'process of socio-material adaptation' (ibid.: 19). The socio-material marker here emphasises the intricate relationship between the social world and its material manifestations – of the built environment or urban infrastructures for example. Urban policy learning invariably plays out across both. Third, underneath these processes of translation and coordination, learning is fundamentally *lived* in the urban context. McFarlane describes this through the notion of dwelling, of inhabiting the world with an emphasis on the body. We learn to live and negotiate the city through everyday practices and sense-making, which both contribute to and challenge other forms of urban knowledge. Urban learning, therefore, emerges as an assemblage of the local and global, the social and material that is inherently performative. It is not the outcome of a ready-made system but continually produced in response to wider urban conditions.

Experimentation fits into this performative, socio-material approach as one possible avenue of urban learning. In its scientific origins, an experiment is described as a practice of knowledge creation that intervenes in an existing system, a specific set-up that generates new knowledge and which enables 'the systematic production of novelty' (Pickstone 2000: 13). But rather than the suggested objective and controlled procedure, experimentation in more recent discussions has been highlighted as a contingent and messy socio-material practice that transcends narrow definitions of spaces, actors and protocols. The experimental laboratory now 'overspills its traditional constitution, inhabiting diverse informational, technological and political environments, changing the nature of experimental sites and experimental subjects' (Davies 2010: 667). This extension brings a variety of potential spaces and stakeholders into experimental efforts, including the lived in places of the city and the diverse actors that navigate daily urban life. It emphasises the role of place in knowledge-making and recognises that knowledge practices are fundamentally situated and emplaced, drawing on a variety of place-features to substantiate knowledge claims and learning. This allows experimentation to be conducted in various, often mundane spaces – from flood defences to people's homes (Marres 2009; Lane *et al*. 2010).

Such openings further 'demand new kinds of responsibility and responsiveness from scientists, politicians, artists and social scientists, as well as citizens' (Davies 2010: 668) – it reframes the actors of experimentation. Citizen science, for example, enlists people not trained as scientists in the production of knowledge in real-life contexts (Irwin 1995), while local conservationists and politicians are engaged in forms of experimental learning through ecological restoration projects (Gross 2010; Lane *et al*. 2010). This also includes recognition of non-human actors in experiments, not only as subjects of testing but active agents within the experimental assemblage (Hinchliffe *et al*. 2005). Experiments need to be negotiated within this assemblage in an ongoing cycle of observation and practical intervention (Gross 2010), emphasising the primary role of daily practices within its knowledge creation.

This extended view on experimentation has been applied to the urban sphere in the context of sustainability transitions, in the form of, for example, strategic niche management (Hoogma *et al*. 2002), climate change experiments (Bulkeley and Castán Broto 2013; Castán Broto and Bulkeley 2013) and urban labs (Evans and Karvonen 2011;, Karvonen and van Heur 2014). The logic of these experiments owes much to the transition and systems innovation approach and its interest in bounded socio-technical experiments (BSTEs). BSTEs are defined as 'an attempt to introduce new technology or service on a scale bounded in space and time' (Brown *et al*. 2003: 292), clearly delineated interventions that are set apart from mainstream dynamics by specific institutional arrangements. As such, they form a particular kind of niche from which innovation may diffuse into the mainstream regime but which also fosters different levels of learning of both a lower-order technical and higher-order social kind. While the former refers to finding solutions to specific problems, the latter emphasises the coming together of different groups to work on shared problem definitions and approaches, rather than their local details (Brown *et al.* 2003). Consequently, these experiments are described as both 'useful' and 'visible' (Evans and Karvonen 2011: 126). They are 'purposive interventions in which there is a more or less explicit attempt to innovate, learn or gain experience' (Bulkeley and Castán Broto 2013: 363), responding to local issues, while being public-facing and designed to attract wider attention. As such, they offer 'a material focus for certain actors to ascribe visions of alternative futures that are also globally recognized' (Karvonen and van Heur 2014: 386). This gives these experiments an important strategic role within urban governance and knowledge-making.

Grassroots initiatives and alternative experiments

Despite this relevance, the strategic character of these experiments and their widespread embeddedness in existing governing or market structures raises the question of whether they are able to substantively challenge and alter existing patterns of urban learning and policy-making. As Gibson-Graham (2008) has argued, particular social or economic structures tend to become dominant not because of any intrinsic features but because of the way they are discursively enacted. Alternatives, similarly, can remain marginal because of a lack of attention. This too applies to urban learning. When implemented within the existing frameworks of urban policy-making, experiments are likely to replicate a number of mostly neoliberal assumptions and structures that are rarely acknowledged. They are filtered through and shaped by existing administrative systems, institutional priorities and underlying values, which impacts on both the kinds of experiments that are implemented, and the way they are enacted and evaluated. Existing dominant systems and pathways are perpetuated, obscuring other possibilities. What is needed, therefore, is a critical approach to urban learning, one which not only improves established paths of knowledge generation but which 'involves questioning and antagonizing existing urban knowledges and formulations, and learning alternative formulations' (McFarlane 2011: 153). Beyond describing the substantive directions of new imaginaries and practices this entails the 'methodo-political task' of identifying 'where they might come from, and how that learning should take place' (ibid.: 154).

The challenge then is to consider a wider range of experimental urban efforts. In their analysis, Bulkeley and Castán Broto (2013: 373) point to 'forms of grassroots experiments [that] co-exist in the city alongside these strategic interventions, raising questions concerning the ability of otherwise marginal actors to use experiments as a means of advancing an alternative politics of climate change'. Yet these more marginal forms of experimentation have received relatively scant attention so far. As Seyfang and Smith (2007) note in their work on grassroots innovations, community action is a 'neglected, but potentially important, site of innovative activity', with examples spanning local currencies, community energy generation and low-impact housing developments (see, for example, Pickerill and Maxey 2009). It is necessary therefore to consider the experimental dimension of such grassroots efforts in more detail.

Like more mainstream innovation niches, they may act as seedbeds for future diffusion, but there are two key aspects of grassroots innovations that distinguish them from purposive experimental interventions (see Seyfang and Smith 2007). First, their operation is open-ended; it emerges organically from ongoing efforts and problem negotiations around existing local social or environmental needs. Second, they show an explicit commitment to alternative values, which allows them to challenge existing formations of learning and production. Here, grassroots innovations intersect with the long history of insurgent, do-it-yourself and guerrilla interventions in the urban sphere, which practise critique through direct intervention and alternative patterns of everyday life. They not only seek to question and disrupt dominant structures, but also to build and test alternative options, often driven by underlying currents of decommodification and the celebration of difference (Iveson 2013). Such alternatives are not automatically experimental in the knowledge-creation sense of the term, and their wider policy-making role is not pre-defined. However, as they actively emphasise their contribution to learning amongst a variety of actors, these grassroots experiments hold much promise for urban change. To understand how particular alternatives come to be urban experiments, it is necessary to tease out the distinctive dynamics and rhythms of their practices and emplacements. The rest of this chapter does this for one notable alternative experiment, the Prinzessinnengarten in Berlin.

More than a garden

The Prinzessinnengarten is an urban garden in the Kreuzberg quarter of Berlin that has come to local and international prominence for its novel take on community gardening and its relevance for urban planning questions. The project began in 2009, when the local initiators leased the previously unused, 6,000 square metre brownfield site on a temporary contract from the city of Berlin. Their aim was to set up an urban agriculture project inspired by the *agricultura urbana* movement in Cuba. But where the Cuban model made use of available community structures to counter a lack of availability of fresh produce, the Berlin version sought to address different issues: a lack of awareness around food production and urban biodiversity, but also a notable lack of local community engagement. Unusually, the garden was set up as a social enterprise, with the view to also challenge established models of community project funding and to create alternative modes of income generation. Since its inception, it has grown steadily in terms of cultivated space, activities and participants, with a core group of 20 to 30 people now working in the garden, as well as over 1,000 volunteers and many more visitors per year (Nomadisch Grün 2015).

It is certainly an unusual and diverse place. A great variety of vegetables, herbs and medicinal plants grow in colourful plastic boxes raised off the ground.

A small café occupies the central part, shaded by a group of black locust trees that are the remnants of the previous abandonment. Other projects sit in between: an old shipping container houses a bicycle workshop, there are beehives and a small shop sells seeds and plants (see Figure 11.1). At times, the garden is a quiet space with only a few people watering or weeding the plants, but it turns into a bustling hub of activity during the upcycling workshops or bee-keeping courses. In fact it is more than a garden; in the words of the garden volunteers:

> After a while you understand that it is so much more [than a garden]. In a way, it is a platform to do things. You can come and just do a sewing workshop, or build a clay oven, or do a bit of gardening. But also, the more you understand this place, the more you see that you can implement all kinds of things here, that you can create things.
>
> (Martina, interview June 2012)

The garden acts as an open space for ideas and projects that are connected to the idea of growing but go beyond its immediate expectations. It connects an emphasis on community work with economic considerations and an educational aim around alternative and sustainable urban living, and it does so through individual projects and ongoing activities that follow a distinctive set of alternative values. It therefore emerges as an intriguing space of grassroots experimentation from its basic thinking and ongoing practices, which will be explored further below.

Figure 11.1 The Prinzessinnengarten in Berlin.

Free space and the art of improvisation

The experimentality of the garden is shaped by an underlying open approach, or 'relaxedness' as Susan and Christina, the initiators of the medicinal plant project, describe it (interview, June 2012). Few fixed rules and goals existed in the garden at its outset; the first activities were only loosely framed by themes of health, integration and biodiversity but mostly followed people's interests and enthusiasm. More structures have developed over the years to manage the growing activities, ranging from the weekly garden group meeting to the priorities created by the demands of hygiene or accounting. Yet these do not determine the garden: they only provide some of the stronger relations in an assemblage that remains fundamentally open-ended. The structures are designed to create 'free spaces', pockets of surprise that allow people to realise projects and ideas, which become experimental openings and which together shape the garden. These openings are both physical and conceptual. Projects like the bees and the bicycle workshop struggled to find a place in the city from which to operate in a non-commercial way until they joined the site. Similarly, the medicinal plant corner resulted from an open invitation by the garden to implement ideas.

But Susan and Christina saw it as more than a physical location for their herbs. They also point to the 'mental free space' that emerges from an absence of judgement, prerequisites and corporate interest. This looseness is summarised by the gardeners when they write, 'The different topics and projects usually enter the garden spontaneously, attracted by the atmosphere of something unfinished and improvised' (Nomadisch Grün 2012: 40). There are perceived and actual openings and connection points that attract ideas, which in conjunction with the existing structures support and direct the diverse garden efforts. This interplay of structures and practised openness provides the adaptive framework needed to make negotiated learning possible in real-world experiments.

These projects themselves are not ready-made either: they are the result of what one volunteer calls the 'art of improvisation' that shapes the garden. This is both an approach and an aesthetic: a willingness to tinker and explore, to play with solutions and accept failures, but also a celebration of recycled or unusual materials and the visibility of traces and processes in the final products. Beer crates are stacked and tied with cable ties to make seats; one of the most-photographed items is a hanging garden made of milk cartons; and the irrigation system is a DIY construction consisting of old hosepipes and flour tanks that looks a little different for each planted row, showing its different stages of development. Underlying these improvised solutions is a set of particular socio-material practices of making that reflect and apply the garden's alternative mindset. Central here is an acknowledgement of material value and agency that directs and enables projects. Materials have a presence in the Prinzessinnengarten; they are stored in a more or less organised manner in various corners, and they require action, for example when new sources of recycling become available. They then act as catalysts and drivers of projects: the irrigation system for example developed from the discovery of cheap recycled tanks and a donation of old hosepipes that could be used to drip-water the beds. This valuation of materials also requires different work rhythms, a willingness to spend time and effort to work with these materials: becoming attuned to their capacities, adapting and combining them, reacting to problems that perhaps would not exist in prefabricated parts.

What is interesting about this process in the garden, and what connects its DIY approach to a wider experimental role, is that this improvisation is publicly cultivated. The garden actively draws attention to these solutions, publishing how-to instructions on postcards, websites and in the garden book. As well as recycling and reuse, practicality is a key concern, as gardener Anton (interview, May 2012) states:

> You need simple, reproducible and functioning solutions. Ones that don't cost much, that anyone can put together … And if something breaks, the material needs to be available again. If a box breaks, I can buy a new box anywhere.

Through its commitment to tinkering and improvisation, the garden sets out a particular material ethic, reminiscent of what Jane Bennett (2001) calls an enchantment with the material world. Seeing everyday materials as valuable, and giving them agency within the development of the garden space, the gardeners cultivate a space 'where enchantment, at least for some of us some of the time, seems to hang out' (ibid.: 169). Here innovative ideas can grow in the city via open and surprising lines, marking the garden as an alternative urban experiment.

Active engagement and tactility

This celebration of openness and improvisation raises the question of how ordinary city-dwellers may come into contact with these alternative practices and approaches, and what their bearings might be on wider issues of alternative and sustainable urban life. The Prinzessinnengarten is based on volunteer participation, ranging from one-off contributions to regulars who take on particular areas of responsibility in the garden, with only a few permanent paid roles. This is reiterated in the definition of membership: after half an hour of work anyone is a member for the day, with discounts on the café and any produce that is harvested. The focus on active involvement feeds into what one volunteer called 'an atmosphere of activity': there is little passive presentation, making sense of the garden requires direct input. Ordering herbal tea from the bar or harvesting produce for example involves cutting one's own herbs and vegetables, with some direction from volunteers and many puzzled looks from the visitors. Instead of prohibitive signs, people are actively encouraged to touch the plants and explore the narrow paths.

The backbone of the garden week is the public work days on Thursdays and Saturdays, during which visitors are invited to join whatever tasks need doing, without prior preparation. This means that within ten minutes of their arrival, visitors might be shovelling compost, mixing tomato soil or repotting mint seedlings. This tactile way of getting to know the space is not limited to a fleeting first encounter. The haptic learning and embodied experience also facilitate a deeper engagement with the space and its concerns. Volunteer Erik (interview, June 2012) describes his longer-term connection to the garden like this:

> Well, really in first place is that I can stick my hands into the soil. That I can work with the soil, that I can feel the soil. That's kind of cleansing and

> healing for me. … And this is why I'm suffering in the city. I can't feel the soil in the city.

This tactile relationship is reflected in his favourite tasks: making new plant beds from different soils, working with the compost, and building a clay oven, all of which are very different from other everyday engagements in the city (Figure 11.2).

Figure 11.2 The clay oven, a product of tactile, embodied, and muddy learning.

The active and haptic experiences also feed into the garden's self-image as an alternative, informal place of learning. Speaking of the construction of the DIY irrigation system, volunteer Julia talks about the need to 'be in the material' to work out exactly how the tanks and hosepipes can be put together. Each step required trials and testing: even if features such as the drip holes immediately make sense, to get the water flow right one needs to actually drill into the rubber and see the results. This is an embodied process, which volunteer Lily (interview, June 2012) describes for the surprisingly difficult task of watering the plastic box beds in the correct way:

> I feel that a lot is about tactile knowledge here. This – I have done it, that's why I know it. It's pretty impossible to just come here and work things out through questions … Like, the watering I learn by doing it and then Anton comes along and taps you on the shoulder and says, 'Hey, don't water like that, I'll show you'. So you have to watch, then you have to go through the process once, or maybe three times. And then maybe Karl comes and says, 'But not like that either …' [laughs] You have to experience a lot of things through the body.

It is a spontaneous kind of learning that arises through the everyday rhythms of the garden, one that depends on chance meetings and conversations, on fitting in with current concerns, and on a willingness to be active and get on with things. This interplay of social and tactile elements makes the garden what Carolan (2007: 1267) calls a 'tactile space': a learning space with both 'a participatory component, which allows individuals to engage in an exchange of knowledge claims through being embedded in social networks, as well as a lived, non-representational component, where individuals physically negotiate their surroundings in an embodied way'. The key value of such a learning space is that it allows big issues to become knowable in everyday mundane, small-scale practices through the affective relations between body and material. The aim is not simply to gain useful skills or to fully comprehend large-scale problems, but to understand connections and to ground issues in personal experience. The advertisement for the gardening day on the garden website indicates this: 'During [the public gardening days] you can get your hands dirty with us and learn useful things about ecological vegetable agriculture in the city, heritage varieties, seed saving, composting methods, seasonality, use and storage etc.' (Nomadisch Grün 2015). The garden's knowledge creation is a process of informal, embodied and often 'chaotic' (Eva, interview June 2012) learning that is continuously negotiated amongst all actors, both human and non-human. It attains its value through its direct application and relevance. This reflects the adaptive and heterogeneous patterns of knowledge-making that characterise open, extended real-world experiments.

An alternative urban green space

The dynamics of openness, improvisation, active engagement and tactile learning mark the Prinzessinnengarten as a unique space in the city, one that is both alternative and experimental. It cultivates alternative approaches and values in its practices, from recycling to social entrepreneurship, and it actively facilitates their testing and sharing in an accessible way. The combination of these practices gives the garden a powerful role within the city as an unusual public space that has an impact on wider urban questions. Although it is operated as a social business with opening hours and a fence, it fulfils many aspects of an urban green public space. It is open to everyone, an island within the bustle of the city that provides a space to relax and de-stress without any prior requirements or the need to consume. But unlike in parks, there is an invitation here to be active. This fosters social contact and communication, and a tactile engagement with the environment that is not possible elsewhere. It creates a unique 'taskscape' (Ingold 1993: 158), a distinctive rhythm that is different from and yet fundamentally linked to the wider city: an *urban* garden. It is this ambiguity of place that makes the space attractive – it rejects the stereotype of a back-to-the-land movement (Halfacree 2007), yet it stands as both a symbolic and practical critical challenge against mainstream urbanism. The garden becomes 'a laboratory for socially and ecologically sustainable forms of urban development' (Clausen 2013: 2), and it attains this role organically, through its emplacement and its ongoing daily practices. As gardener Nils (interview, June 2012) explains:

> This garden was not created as an architectural, urban planning, city development project. But because we are doing a garden in the city, on a brownfield site that is owned by the city, there is an unintentional political role.

Where other urban experiments are designed with a strategic role in mind, driven by particular governance objectives, alternative experiments like the Prinzessinnengarten grow into this role over time. This gives them a greater local grounding without being static. They gain increasing depth and resilience through their open dynamics, multiple inputs and improvised approaches. This in turn enables them to become active in the face of diverse pressures, carving out a space from which to have an input into ongoing political discussions. These includes immediate concerns such as negotiating the lease agreements for the space and the current planning processes around brownfield sites in Berlin (see Initiative Stadt Neudenken 2011). But it also addresses long-term questions of city planning in the face of climate, social and demographic change. The garden acts as a 'showroom for questions' (Nils, interview June 2012), as a tangible focal point for discussions, and a lived site of knowledge production.

Here, the garden reflects the promise of an alternative mode of experimentation, one that is emergent and applies the method of experimental testing with a critical angle. As Bulkeley and Castán Broto (2013: 367–368) add to their analysis of strategic examples, '(r)ather than creating protected spaces through which innovation can be fostered and system change developed, experiments could provide grist in the urban mill, creating conflict, sparking controversy, offering the basis for contested new regimes of practice'. The Prinzessinnengarten is not a protected space, as the diverse debates around its future show, nor is it a replicable blueprint. Although many gardens have been inspired by its success, both within Berlin and beyond, its primary aim is not to spread urban agriculture through simple diffusion or transfer. Rather, it seeks to raise questions and provide the tools with which answers to a variety of questions can be worked out. Given an understanding of urban learning as a complex, assembled process that includes a local, experiential and practised component (McFarlane 2011), alternative experiments take up a distinct and valuable role in processes of urban change. They provide the spaces and milieus (Longhurst 2013) where new, innovative practices are shaped, performed and tested. The resulting formations are then adapted to feed into other domains of the city, through personal, experiential learning but also through new inputs into wider policy discussions. The emplacement of these grassroots experiments allows for a more organic emergence of ideas, solutions and responses that take their starting point from local needs. Alternative experiments offer a different contribution to processes of urban learning, one that provides a voice to otherwise marginal actors and ideas, challenging established processes for a more diverse, and perhaps more hopeful, urban future.

Note

1 The dynamics by which alternative communities in occupied spaces implement their visions of self-governed, low-impact living, and how they emerge as experiments, is considered in more detail in my PhD thesis (Wendler 2014).

References

Bennett, J. (2001). *The Enchantment of Modern Life*. Princeton, NJ: Princeton University Press.

Brown, H.S., Vergragt, P., Green, K. and Berchicci, L. (2003). Learning for sustainability transition through bounded socio-technical experiments in personal mobility. *Technology Analysis and Strategic Management*, 15: 291–315.

Bulkeley, H. and Castán Broto, V. (2013). Government by experiment? Global cities and the governing of climate change. *Transactions of the Institute of British Geographers*, 38: 361–375.

Can Masdeu (2014). *Vall de Canmasdeu – FAQ*. [Online]. Available: http://www.canmasdeu.net/la-vall/faq/ [27 June 2014].

Carolan, M.S. (2007). Introducing the concept of tactile space: creating lasting social and environmental commitments. *Geoforum*, 38: 1264–1275.

Castán Broto, V. and Bulkeley, H. (2013). A survey of urban climate change experiments in 100 cities. *Global Environmental Change*, 23: 92–102.

Christiania (2004). *Christiania Guide*. [Online]. Available: http://www.christiania.org/wp-content/uploads/2013/02/Guideeng2.pdf [16 January 2015].

Clausen, M. (2013). *Community Based Urban Agriculture and Resilient Food Systems from a Bottom-up Point of View*. [Online]. Available: http://prinzessinnengarten.net/wp-content/uploads/2013/11/Prinzessinnengarten_5AesopFoodPlanning.pdf [12 January 2015].

Cooper, D. (2014). *Everyday Utopias: The Conceptual Life of Promising Spaces*. Durham, NC: Duke University Press.

Davies, G. (2010). Where do experiments end? *Geoforum*, 41: 667–670.

Evans, J. and Karvonen, A. (2011). Living laboratories for sustainability: exploring the politics and epistemology of urban transition. In Bulkeley, H., Castán Broto, V., Hodson, M. and Marvin, S. (eds), *Cities and Low Carbon Transitions*. London: Routledge, 126–141.

ExperimentDays. (2014). *ExperimentDays 14*. [Online]. Available: http://experimentdays.de/2014 [24 June 2014].

Gibson-Graham, J.K. (2008). Diverse economies: performative practices for 'other worlds'. *Progress in Human Geography*, 32: 613–632.

Gross, M. (2010). *Ignorance and Surprise: Science, Society, and Ecological Design*. Cambridge, MA: MIT Press.

Halfacree, K. (2007). Back-to-the-land in the twenty-first century – making connections with rurality. *Tijdschrift voor Economische en Sociale Geografie*, 98: 3–8.

Hinchliffe, S., Kearnes, M.B., Degen, M. and Whatmore, S. (2005). Urban wild things: a cosmopolitical experiment. *Environment and Planning D: Society and Space*, 23: 643–658.

Hoogma, R., Kemp, R., Schot, J. and Truffer, B. (2002). *Experimenting for Sustainable Transport: The Approach of Strategic Niche Management*. London: Spon Press.

Ingold, T. (1993). The temporality of the landscape. *World Archaeology*, 25(2): 152-174.

Initiative Stadt Neudenken. (2011). *Initiative Stadt Neudenken*. [Online]. Available: http://stadt-neudenken.tumblr.com/ [5 September 2014].

Irwin, A. (1995). *Citizen Science: A Study of People, Expertise and Sustainable Development*. London: Routledge.

Iveson, K. (2013). Cities within the city: do-it-yourself urbanism and the right to the city. *International Journal of Urban and Regional Research*, 37: 941–956.

Karvonen, A. and van Heur, B. (2014). Urban laboratories: experiments in reworking cities. *International Journal of Urban and Regional Research*, 38: 379–392.

Lane, S.N., Odoni, N., Landström, C., Whatmore, S.J., Ward, N. and Bradley, S. (2010). Doing flood risk science differently : an experiment in radical scientific method. *Transactions of the Institute of British Geographers*, 36: 15–36.

Longhurst, N. (2013). The emergence of an alternative milieu: conceptualising the nature of alternative places. *Environment and Planning A*, 45: 2100–2119.

McCann, E. and Ward, K. (eds) (2011). *Mobile Urbanism: Cities and Policymaking in the Global Age*. Minneapolis: University of Minnesota Press.

McFarlane, C. (2011). *Learning the City: Knowledge and Translocal Assemblage*. Oxford: Wiley Blackwell.

Marres, N. (2009). Testing powers of engagement: green living experiments, the ontological turn and the undoability of involvement. *European Journal of Social Theory*, 12: 117–133.

Nomadisch Grün (ed.) (2012). *Prinzessinnengärten. Anders gärtnern in der Stadt*. Berlin: Dumont.

Nomadisch Grün (2015). *Prinzessinnengärten*. [Online]. Available: http://prinzessinnengarten.net/ [10 May 2015].

Peck, J. (2011). Geographies of policy: from transfer-diffusion to mobility-mutation. *Progress in Human Geography*, 35: 773–797.

Pickerill, J. and Maxey, L. (2009). Geographies of sustainability : low impact developments and radical spaces of innovation. *Geography Compass*, 3/4: 1515–1539.

Pickstone, J.V. (2000). *Ways of Knowing: A New History of Science, Technology and Medicine*. Manchester: Manchester University Press.

Seyfang, G. and Smith, A. (2007). Grassroots innovations for sustainable development: towards a new research and policy agenda. *Environmental Politics*, 16: 584–603.

Wendler, J. (2014). *Experimental Urbanism: Grassroots Alternatives as Spaces of Learning and Innovation in the City*. PhD Thesis, University of Manchester.

12 Living labs

Users, citizens and transitions

Gabriele Schliwa and Kes McCormick

Introduction: from the city as laboratory to living labs

Real-life environments have been used and framed as natural laboratories in which to study and develop new knowledge and understandings of human behaviour since the start of the last century (if not before). A hundred years ago, Park (1915: 612) developed frameworks for analysing processes of social change within cities:

> The city, in short, shows the good and evil in human nature in excess. It is this fact, perhaps, more than any other which justifies the view that would make of the *city a laboratory* or clinic in which human nature and social processes may be most conveniently and profitably studied [emphasis added].

Likewise, urban researchers have been studying the phenomenon of urban experimentation for a long time (Bulkeley and Castán Broto 2013; Karvonen *et al.* 2014). Over the last decade, the city has been increasingly cast as a laboratory for the study of sustainable development (Evans and Karvonen 2011). In particular, an increasing number of institutions call themselves a 'living lab', demonstrating the level of interest in this concept from many different stakeholders, such as universities, science parks, business and local governments. Living labs have an appeal as they can suggest rigour and innovation, and in some instances become almost a model for urban development (Evans and Karvonen 2014).

While it was not the first time the term living lab had been used (see, for example, Abowd *et al.* 2000), a white paper published in the *Annual Progress Report of the Center for User–System Interaction* described living labs to be 'pivotal for user–system interaction research in the next decade' (Markopoulos and Rauterberg 2001: 65). Furthermore, the report states the expectations for the concept:

> The living lab is a planned infrastructure that will provide an *experimental platform* for future home-related technologies. We plan partnerships for developing the required technologies, and for testing research concepts or novel products in an *'ecologically valid'* manner [emphasis added].

Since 2001, the living lab approach has been increasingly tested and applied in the corporate ICT sector (Følstad 2008). In 2003, William J. Mitchell from the

Massachusetts Institute of Technology MediaLab, described their PlaceLab as 'a new kind of scientific instrument – a "microscope" to carefully study people and their interaction with new technologies in a living environment,' and further elaborates 'with the extraordinary pace of technological development, and with the potential for innovations to dramatically improve people's lives, it is essential that researchers better understand how to design systems that people will want to bring into their homes' (MIT News 2003). Mitchell is often credited to be the creator of the concept (Eriksson *et al.* 2005; Mensink *et al.* 2010; Almirall and Wareham 2011; Schuurman *et al.* 2011; Pallot *et al.* 2013).

In 2006, the European Network of Living Labs (ENoLL) was founded as the international federation of benchmarked living labs covering 26 European countries and 8 additional countries in Africa, North and South America, and Asia (ENoLL 2013; European Commission 2013a). The sectors in which the ENoLL community operates can be characterised in several thematic domains, which in 2011 were represented as: Creative Industries and E-learning (39 per cent); Ambient Assisted Living, E-Health and Sports (29 per cent); Intelligent Energy, Smart Grid and Sustainable Building (13 per cent); and Transport, Logistics and Automotive (9 per cent) (Alcotra Innovation 2011). In its *Open Innovation 2.0 Yearbook 2013*, the European Commission (2013b) describes living labs as an emerging innovation platform to bring forward the quadruple helix model of a so-called Public–Private–People–Partnership (PPPP) where citizens are intended to have a strong influence on the innovation process.

Leminen *et al.* (2012) proposed four types of living labs: utiliser-driven, enabler-driven, provider-driven, and user-driven. The types are defined by the actor that plays the most dominant role in the initial phase or later acts as the principal promoter of innovation activities. They differ in terms of activities, structure, organisation, and coordination. *Utilisers* are often companies using a living lab for product–service system development for the private sector, *enablers* are often but not exclusively local governments representing the public sector, *providers* are mainly research institutions and universities that in most cases host the living lab and, finally, *users* are the people involved in the projects represented according to the European Commission's PPPP definition. Major innovation research funding schemes such as the EU Seventh Framework Programme for Research (FP7) (2007–2013) as well as the current Horizon 2020 Research and Innovation programme (2014–2020), which with nearly €80 billion of funding is the biggest funding scheme ever, explicitly encourage the involvement of end users as well as living lab innovation research infrastructures in their funding calls (Schliwa 2013; Voytenko *et al.* 2015).

More recent literature and projects have shown that the living lab approach has been adopted as a tool in urban governance and sustainability research (Evans and Karvonen 2011, 2014; König 2013; McCormick *et al.* 2013; Nevens *et al.* 2013; Schneidewind and Scheck 2013). In this context, cities represent an ideal arena for experimentation and real-world experiments are acknowledged as a way to generate knowledge about the emergence, development and diffusion of system innovation for sustainable urban development (Schneidewind and Scheck 2013; Evans and Karvonen 2014). Hence, the living lab approach used in sustainability research is increasingly linked to a broader territorial perspective to use places such as city districts and university campuses as arenas for real-life experiments. Living labs that operate on an urban territory in the pursuit of sustainability goals are increasingly being conceptualised and defined as so-called urban living labs (Juujärvi and Pesso 2013; Nevens *et al.* 2013; Schliwa 2013; Wallin 2014; Voytenko *et al.* 2015).

This chapter explores how the living lab approach has been transferred from the private to the public sector and has emerged as a research infrastructure and governance tool for urban transitions. It expands on existing living lab typologies to enable practitioners, policy makers and researchers, as well as citizens engaging with them, to develop current understandings of how living labs in the urban context are conceived, how they function and ultimately, how they achieve their desired impacts.

Distinguishing between users and citizens

To date, there is no clear definition of the term 'living lab', as the function varies between different host institutions as well as funding bodies. The rhetoric around living labs seems to adapt quickly to popular themes. Terms describing their characteristics such as 'user', 'citizen' or 'people' as well as 'centric' or 'driven' often appear to be used interchangeably and the lines blur between newly established categories. For example, ENoLL describes living labs most commonly as '*user-driven* innovation environments where users and producers co-create innovation in a trusted, open ecosystem that enables business and societal innovation' (Eskelinen *et al.* 2015: 12). However, their recent publication together with the World Bank (based on a joint Memorandum of Understanding from 2012) is titled *Citizen-Driven Innovation* and in this report, the term 'citizen' outnumbers the term 'user' despite a focus on digital product and service development (World Bank 2012; Eskelinen *et al.* 2015).

Since its emergence in academic research, the topic of living labs has become very diverse, which makes it difficult to analyse cases and challenging to effectively categorise them (Moor *et al.* 2010; Farrall 2012). Nevens *et al.* (2013) state that their urban transition lab concept is inspired by the living lab concept and tailored to urban settings on a local level. As Juujärvi and Pesso (2013) observe, studies equivalent to regional living lab activities have previously been framed as regional innovation networks (Melkas and Harmaakorpi 2008; Kallio *et al.* 2010) or participatory urban planning (Horelli *et al.* 2013). Several definitions exist, as shown in Table 12.1. The examples show how various approaches differ but use related terminologies to describe the living lab approach in an urban context. As Juujärvi and Pesso (2013: 23) further explain, 'an urban living lab can be seen as a special type of regional innovation network that puts emphasis on residents and their communities as users (i.e. ordinary people who want to solve their real-life problems)'.

Table 12.1 Definitions for urban living labs

Organisation	*Definition*	*Keywords*
Eurbanlab (Network)	'An urban living lab is a user-centred, open innovation ecosystem, operating in a territorial context, integrating concurrent research and innovation processes within a private–public–people partnership. Urban living labs are breeding grounds for innovation and play a crucial role in the process of transitioning towards low carbon resilient cities.' (Eurbanlab 2012).	User-centred, open innovation, territorial context, PPPP, transitioning, low-carbon cities
Nevens *et al.* (Academia)	'We consider an Urban Transition Lab as the locus within a city where (global) persistent problems are translated to the specific characteristics of the city and where multiple transitions interact across domains, shift scales of operation and impact multiple domains simultaneously (e.g. energy, mobility, built environment, food, ecosystems). It is a hybrid, flexible and transdisciplinary platform that provides space and time for learning, reflection and development of alternative solutions that are not self-evident in a regime context.' (Nevens et al. 2013).	Locus within a city, specific, multiple, learning, reflection
JPI Urban Europe (Funder)	'It is a forum for innovation, applied to the development of new products, systems, services, and processes, employing working methods to integrate people into the entire development process as users and co-creators, to explore, examine, experiment, test and evaluate new ideas, scenarios, processes, systems, concepts and creative solutions in complex and real contexts.' (JPI Eurban Europe 2012).	Forum for innovation, integrate people, complex real-life context

Sources: Pallot *et al.* 2010; Eurbanlab 2012; JPI Urban Europe 2013; Nevens *et al.* 2013.

This chapter refers to the term 'living lab' in two ways: as an *approach* putting users at the centre of innovation processes and as an *arena* for experimentation (Ståhlbröst 2012). Evans and Karvonen (2014) identify three core characteristics of living labs that are significant for both the approach and the arena interpretations: (1) they are a geographically and/or institutionally *bounded space*; (2) they conduct *intentional experiments;* and, (3) they display *iterative learning*. These characteristics combine to create the foundations for living labs as well as urban living labs.

The term *arena* addresses the first core characteristic that living labs are a *geographically* (e.g. a building, a university campus or an entire city) or *institutionally* (e.g. involving key actors) bounded space. The term *approach* includes the second and third core characteristics. It describes the process of intentional experiments for socio-technical innovation together with the four stakeholder groups (citizens, academia, business and the public sector). Iterative learning means that experiments are conducted, monitored, and conducted again with improvements from the previous round to generate useful knowledge in a real-life setting. These iterative learning loops aim to improve future products and services as well as societal and technical structures within the urban environment (Ståhlbröst and Holst 2013; Evans and Karvonen 2014).

Our investigation reveals two research streams in the living lab literature. Research streams describe a series of related papers on one topic that increasingly deepens the knowledge of the area under investigation. In the 'optimal' case, this development starts with theory papers, continues with qualitative and case study research, subsequently conducts quantitative research, and finally may lead to policy papers (Yin 2003; Peng 2010). What all living labs have in common is the introduction of users as the new key stakeholder in the experimentation process. User involvement can position the user as the main creator, in the case of lead users (von Hippel 1986), as co-creators in creative practices such as design thinking (Brown 2008) or as passive subjects such as market validation exercises. Living labs are situated in the middle ground of user involvement (Almirall *et al.* 2012).

This chapter defines human-centric design and innovation as the underlying default for both living lab research streams, and the suffix 'driven' identifies the host institution. The first research stream describes living labs that mostly target ICT product–service system development as *user-centric living labs* (UCLL) with people in their role as users. The second research stream concerns living labs that seek sustainable urban transitions. These are therefore defined as *citizen-centric living labs* (CCLL) with people in their role as citizens and not necessarily as users. While it is not feasible to draw a clear line between these categories, as both operate with the same approach and the description of people as citizens or users is not mutually exclusive, there is a clear distinction in practice and meaning. The CCLL innovation process goes beyond the development of product–service systems and seeks to tackle sustainability challenges with citizens and academia as central actors involved in solving urban problems.

While research on living labs is increasing, there is still little systematic research on citizen-centric living labs within their urban context (Juujärvi and Pesso 2013). Existing typologies are based on case studies analysing the concept from a public–private sector perspective, where UCLL approaches are applied as a form of open-innovation network creating competences and competitive advantages for the development of new business strategies and opportunities. A considerable number of these publications are published in *Technology Innovation Management Review*, user handbooks or policy documents with a scholarly community of business, innovation and technology actors (Pascu and van Lieshout 2009; Leminen *et al.* 2012; Schuurman *et al.* 2012). This chapter contributes to the CCLL research stream that has been further labelled as urban living labs. The scholarly community emerges from social science with a focus on sustainability transitions, urban governance and experimentation (Evans and Karvonen 2011, 2014; Bulkeley and Castán Broto 2013; McCormick *et al.* 2013; Nevens *et al.* 2013; Voytenko *et al.* 2015).

Table 12.2 provides an overview of the characteristics of user-centric and citizen-centric living labs. Comparing the key characteristics, user-centric living labs are more closely defined and restricted with respect to their urban area, stakeholder involvement, time horizon and the type of products tested. The study by Juujärvi and Pesso (2013: 22) distinguishes amongst at least three types of urban living labs. The first category refers to urban areas as technology-assisted research environments 'in which users give feedback on products and services',

Table 12.2 Characteristics of living lab research streams

Characteristics	*User-centric living labs (UCLL)*	*Citizen-centric living labs (CCLL)*
Geographically bounded space	Small-scale living lab (e.g. an existing home or a constructed apartment up to entire city since UCLL are more recently associated with smart cities).	Larger scale living lab (e.g. university campus, a city or region as well as living labs as a virtual space that facilitates a collaborative process).
Time-wise bounded space	Short-term; project duration can range from a few weeks up to several years depending on funding and finance available.	Mid-term to long-term depending on funding available. However, often not clearly defined as the aim is to expand the activities within the city for permanent changes of the urban and social fabric.
Institutionally bounded space	Mostly utiliser-driven, more selective involvement of key stakeholders and users to have a representative combination of suitable actors; mutual benefits of involving the local government.	Selective involvement of key stakeholders, beyond that everyone is considered to bring an added value and create a critical mass; engagement of local government necessary in order to have the license to operate and expand on the urban territory.
Intentional experimentation	Development of product and service system innovations (e.g. ICT-based home energy management systems).	Experimentation including product and service system development as well as new forms of collaboration, employment and education.
Knowledge generation	Focus on knowledge generation within a more controlled and monitored real-life setting.	Focus on knowledge application within a less controlled and monitored real-life setting.
Example networks	ENoLL, Human Smart Cities Network, Open & Agile Smart Cities.	International Sustainable Campus Network (ISCN), JPI Eurban Europe, Eurbanlab.

Sources: Developed based on case studies, Evans and Karvonen 2014 and Leminen *et al.* 2012.

the second is described as 'users can co-create (…) local services', while the third refers to urban planning and vision-making 'with the engagement of citizens'. Implicitly, this makes a distinction between users and citizens.

Catalysing transitions

Living labs have favourable characteristics for catalysing transitions across institutional and geographical boundaries. While the body of research on living labs in general is increasing, there is little systematic research on living labs within an urban context and with citizens as users. Therefore, data were collected through a literature review, expert interviews and case study research on five geographically distinct living labs in Europe: the Urban Living Lab in France (ULL-VSQY) (Yvelines Conseil Général 2010; Garnier 2013 ULL-VSQY 2013;); the Città Studi Campus Sostenibile (CS) in Italy (Periphèria 2011; Campus Sostenibile 2012; Concilio 2013); the Botnia Living Lab (BLL) in Sweden (Bergvall-Kareborn *et al.* 2009; Alvsilver 2013; Braškus 2013; Einarsson 2013; Holst 2013; Krogstie *et al.* 2013; Ståhlbröst 2013); Low Impact Living Affordable Community (LILAC) in the UK (Chatterton 2014a, 2014b; Sherwood 2014) and the SusLabNRW project in Germany (Liedtke *et al.* 2012; SusLabNWE 2012; Baedeker 2013).

Table 12.3 provides an overview of the key characteristics of the five case study living labs and their projects. The classification reflects the four types of living lab drivers suggested by Leminen *et al.* (2012) to define the institutional boundary, as well as the territorial scale to define the geographical boundary. Further distinctive characteristics such as project goals, project duration, partners and finance are provided. All of the selected living lab case studies have a strong link with academia, either through their initiating institution or their proximity. With regards to the geographically bounded space, living labs range from small-scale (e.g. an apartment or a building) to large-scale (e.g. region) projects. For example, the term sustainable living lab is introduced by Liedtke *et al.* (2012) as a research approach explicitly addressing sustainable homes as an arena of major relevance.

The way living labs are strategically set up and tactically expand their network based on the driving institution has major implications for whether their intentional experiments facilitate urban transitions. This depends on the type of institution driving the activities and whether the core team succeeds in engaging with further actors through joint experimentation, as well as the impact demonstration of pilot projects. Table 12.4 visualises this relationship and suggests a scale ranging from low (+) to high (+++) to indicate the probability of a given combination to achieve transformative impact within its institutional and geographical boundaries. This suggestion is subject to discussion and requires further research.

A provider-driven living lab can for example expand its activities outside the university campus when local authorities provide support and the licence to operate in other visible parts of the city. This is represented by the CS case study, where the provider-driven urban living lab inspired the Municipality of Milan to further engage within the experiments and provided access to urban areas outside the university campus (Concilio 2013). As Leminen and colleagues (2012) state, user-driven living labs

Table 12.3 Key characteristics of living lab case studies

Name	*Botnia Living Lab (BLL)*	*Low Impact Living Affordable Community (LILAC)*	*Città Studi Campus Sostenibile (CS)*	*Urban Living Lab (ULL-VSQY)*	*SusLabNRW (SusLabNRW)*
Location	Lulea, Sweden	Leeds, UK	Milan, Italy	Versailles, France	Bottrop, Germany
Territory – Geographically bounded space	Building, also university campus, inter-urban/international	District, also building	University campus, also city	Region, also university campus, city, inter-urban/international	Inter-urban/ international, also building
Driver – Institutionally bounded space	Utiliser-driven	User-driven	Provider-driven	Enabler-driven	Utiliser-driven, enabler-driven
Living Lab host	Lulea Technical University	LILAC community	Politecnico de Milano	Fondaterra	Wuppertal Institute
Project goals and fields of interest	Development of living labs methodology, energy efficiency projects. Energy saving solutions for households and transportation, development of living labs methodology.	Co-housing project that develops a mutually supportive low-impact living community at affordable costs. Locally sourced building material (straw) and food, passive solar buildings, design principles, sharing lifestyles and interaction.	Urban model in Milan for life quality and environmental sustainability. To test innovations developed by scientific research. To promote life style transformation and more liveable spaces.	Share knowledge and experiences relating to territories innovation strategies and their implementation. Low carbon city (energy efficiency and urban planning), high quality of life.	Development of measures to increase energy and resource efficiency in buildings. User integrated processes for energy and resource efficient product service innovations.

Living Lab/Project duration	Permanent, started in 2000. Project (limited). Duration of 'SmartIES' project: 2010–2012.	Long-term, permanent. Running since 2013 after many years of planning and construction.	Project (open end). Creation of launch team in 2011. Official launch of the project in 2011.	Permanent, official living lab status by ENoLL since 2011, previous projects under Fondaterra	Project (limited). Duration of 'SusLabNRW' project: 2012–2015.
Network memberships	ENoLL, FIRE	Member of LILAC	INSC, Periphèria, ENoLL	ENoLL, Eurbanlab	SusLabNWE
Partners	NTNU, Iceconsult Iceland, Sunrise Valley, Vinnova, Lulea Energi, Lulea University, and many others	Community of 35 adults and 10 children forming a Mutual Home Ownership Society	PoliMI, UniMI, City of Milan, private companies, and many others	UVSQ, Saint-Quentin-en-Yvelines and Versailles Grand Parc, Renault, Peugeot, SUEZ, and many others	Hochschule Ruhr West, Innovation City Bottrop, Fraunhofer InHaus,Dt. Telekom, Evonik, Vaillant and many others
Project finance	Public, e.g. Vinnova, EU FP7 and others	Community run leasing scheme, Mutual Home Ownership Society, LILAC Equity Fund	Public, little private, e.g. EU projects	Public and private, e.g. EU projects, private companies	Public, e.g. Interreg, national funding

Sources: Developed based on case studies, Evans and Karvonen 2014 and Leminen *et al*. 2012.

Table 12.4 Impact between institutional boundaries and geographical scales

Institutional boundary	*Building*	*District*	*University campus*	*City*	*Region*	*Inter-urban/ international*
Utiliser-driven (e.g. company)	+++ (BLL)	++	+	++	+	++
Enabler-driven (e.g. municipality)	++	++	++	+++	+++ (ULL)	++
Provider-driven (e.g. university)	++	++	+++ (CS)	+++	+++	+++ (SusLabNRW)
User-driven (e.g. citizen)	++	+++ (LILAC)	++	+	+	+

Sources: Developed based on case studies, Evans and Karvonen 2014 and Leminen *et al.* 2012.

are rare but powerful. This is represented by the LILAC case study, where citizens built a mutually supportive community (here categorised as district) with high transformative impact within the community but this is unlikely to expand to a regional scale when only managed by their users. The broader impact of such an urban experiment can be achieved through a 'niche break out' (small scale replication in other geographical contexts) rather than a 'niche breakthrough' (expansion from district level to regional level) (Chatterton 2014a).

Measuring and reporting impacts play a key role for the learning effect and further development of living labs. To evaluate the impacts of living lab projects, outputs are linked to three types of impact – direct, indirect and diffuse (Loorbach 2013). The indicators and benchmarks used for the evaluation of a living lab project influence the way and extent to which they contribute to new sustainable structures and processes for urban transitions.

Direct Impact is created within the scope of a living lab project. It creates the most tangible outcome and is measured from an economic perspective (e.g. costs of the product, job creation, reduction of bills, willingness to pay and life cycle costs), from an ecological perspective (e.g. resource efficiency, energy efficiency, greenhouse gas emissions, carbon footprint) and from a social (or so-called user) perspective (e.g. acceptance of technologies, high quality of life, number of participants involved in a project).

Indirect Impact results in follow-up activities that are beyond the scope of the project but inspired by it. For example, indirect impact occurs in the form of adjusted policy regulations. In practice, this is enforced by interconnecting environmental, social and economic sustainability (e.g. targeting employment and education within the experiment portfolio). In this way, living labs can create a bottom-up push but also a top-down policy pull to legitimise operations to influence local regulatory frameworks. It is important that innovations and activities created in a living lab do not remain in the academic sphere but are transferred into the market and society.

Diffuse Impact can be considered the most important outcome for a successful transition process. The diffuse impact refers to a change of cultural and normative values within a society, which may influence the perception of sustainability problems and the design of urban infrastructures. It is when 'people don't realise anymore that they make the change' (Loorbach 2013). This often lies beyond the scope of the project, it is difficult to measure and it can often only be monitored with a significant time delay.

The most beneficial impact to drive transitions is at the same time the least tangible one. The mere consideration of the direct impact depends on stakeholder interests and how they measure their individual success. Lobbying and structural power from the public or private stakeholder side can hence lead to a deviation in the long-term sustainability vision. Living labs need to consider their impact beyond the scope of a single, short-term project to successfully contribute to transition processes.

An unexpected finding of the research was that funding agencies and schemes, such as FP7 (European Commission 2013a) as well as the current Horizon 2020 framework, increasingly favour projects with user-engagement or even explicitly suggest the living lab approach in their 'call for proposal' documents. Also, living lab hosts explained that after their organisations adopted the living lab banner, it was easier for them to attract funding from European funders.

Despite the fact that many scholars associate the concept with the Massachusetts Institute of Technology, recent literature reviews identify the concept as a largely European phenomenon with only a few initiatives in the USA and Australia (Schuurman 2015). Hence, a possible explanation for the increased attention to the living lab approach in the last decade is that user-engagement has become a requirement to secure European funding (Talwar *et al.* 2011; Schliwa 2013). User involvement takes different forms in innovation practice (see, for example, von Hippel 1986). Design thinking fulfils the full spectrum of innovation activities with a human-centred design ethic (Brown 2008). What is observed here is the financially incentivised introduction of a design thinking methodology into urban governance.

Interestingly, in parallel with the increasing attention the living labs concept has gained in innovation and urban research, the smart city concept emerged in a similar manner. An early and widely cited paper on this topic, 'The vision of the smart city' (Bowerman *et al.* 2000), described an initiative led by the Brookhaven National Laboratory. Therein, the authors state that the smart cities concept had been at a planning stage since late 1998 and received its first funding in early 2000. The authors further elaborate on the need to grow and sustain a smart city vision 'through the establishment of a sustaining funding engine' (Bowerman *et al.* 2000: 5). Since these early days, the use of the term 'smart city' has increased exponentially, even eclipsing 'sustainable city' by 2012 (de Jong *et al.* 2015). While a major body of academic literature criticises the smart city concept for its overly technocentric and undemocratic approach to urban governance (Greenfield 2013; Kitchin 2014; Viitanen and Kingston 2014; Luque-Ayala and Marvin 2015), policy guidebooks such as the recent joint publication by ENoLL and the World Bank 'explore the concept of smart cities through a lens that promotes citizens as the driving force of urban innovation', promoting living labs as complementary initiatives (Eskelinen *et al.* 2015).

Conclusions

Living labs can be defined as a physical arena as well as a collaborative approach in which different stakeholders have space to experiment, co-create and test innovation in real-life environments defined by their institutional and geographical boundaries. National and international networks are vital for living labs to share and spread knowledge, to jointly develop methodologies and evaluation indicators, to increase their visibility and to attract funding. As such, individual living labs as well as their networks are gaining more structural and operational power. The partnerships between researchers, citizens, companies and local governments within living labs create beneficial preconditions to connect innovations with the market and society, and to advance transitions.

The major finding of this exploratory research is that the living lab approach has developed in two research streams. The first is living labs with a focus on product–service system development. These can be categorised as *user-centric* living labs and this has been the general definition and default in early studies. This implies that the living lab host organisation perceives the citizen as a user in the often digitally enabled product–service system innovation that is being developed within the institutional boundaries of the living lab. In practice, this stream is becoming increasingly used and associated with the smart city concept. A second research stream emerged around the term urban living lab as a *citizen-centric* perspective. Urban living labs with a citizen-centric approach develop their activities within an urban area with the goal of inspiring citizens beyond their current geographical boundaries to achieve transformative impact that can drive sustainable urban development. In contrast to the association of user-centric living labs with the smart city concept, citizen-centric living labs suit sustainable city strategies.

Distinguishing between user-centric and citizen-centric living labs can be important in understanding how different types of living labs potentially contribute to sustainable urban development. Both types of living labs provide physical and financial space for experimentation that may not have taken place otherwise. The knowledge generated in citizen-centric urban living labs is a key contribution to advancing sustainable urban transitions. Unlike user-centric living labs, it is of minor importance if a technology developed in a citizen-centric living lab turns out to be a success or not. Despite several challenges, promising living lab characteristics are found to advance sustainable urban transitions in practice. That said, the living lab approach is one of many tools to drive innovation and requires further research to assess its impact. These initial observations require further inquiry into the design of funding schemes, the history of the smart city and the living lab concepts as a corporate method within urban governance, as well as the adoption of research concepts in different urban sectors.

References

Abowd, G.D., Atkeson, C.G., Bobick, A.F., Essa, I. A., MacIntyre, B., Mynatt, E.D. and Starner, T.E. (2000). Living laboratories: the future computing environments group at the Georgia Institute of Technology. In *Proceedings of Conference on Human Factors in Computing Systems (CHI '00): Extended Abstracts on Human Factors in Computing Systems*. New York: ACM Press, 215–216.

Alcotra Innovation (2011). *Best Practices Database for Living Labs: Overview of the Living Lab approach. Alcotra Innovation Project*. [Online]. Available: http://www.alcotra-innovation.eu/progetto/doc/Best.pdf [18 September 2015].

Almirall, E. and Wareham, J. (2011). Living labs: arbiters of mid- and ground-level innovation. *Technology Analysis and Strategic Management*, 23: 87–102.

Almirall, E., Lee, M. and Wareham, J. (2012). Mapping living labs in the landscape of innovation methodologies. *Technology Innovation Management Review*, 2: 12–18.

Alvsilver, A. (2013). Interview with Alexander Alvsilver, 17 July 2013. Vinnova (Swedish Agency for Innovation Systems).

Baedeker, C. (2013). Interview with Carolin Baedeker, SusLabNWR project.

Bergvall-Kareborn, B., Holst, M. and Ståhlbröst, A. (2009). Concept design with a living lab approach. In *Proceedings of the 42nd Hawaii International Conference on Systems Science*. Washington, DC: IEEE Computer Society, 1–10.

Bowerman, B., Baverman, J., Taylor, J., Todosow, H. and von Wimmersperg, U. (2000). The vision of a smart city. Paper presented at the *2nd International Life Extension Technology Workshop*. Paris, 28 September 2000.

Braškus, L. (2013). Interview with Laurynas Braškus, Sunrise Valley Science and Technology Park.

Brown, T. (2008). Design thinking. *Harvard Business Review*, 86: 84–92.

Bulkeley, H. and Castán Broto, V. (2013). Government by experiment? Global cities and the governing of climate change. *Transactions of the Institute of British Geographers*, 38: 361–375.

Campus Sostenibile (2012). *ISCN-GULF Sustainable Campus Charter Report 2012*. [Electronic]. Available: http://www.campus-sostenibile.polimi.it/iscn1 [18 September 2015].

Chatterton, P. (2014a). LILAC. Presented at the workshop *Universities, Cities and Transformation: Practices of Cultural Intermediation and Expectations of Knowledge*, 1–2 September 2014. Manchester.

Chatterton, P. (2014b). *Low Impact Living: A Field Guide to Ecological, Affordable Community Building*. New York: Routledge.

Concilio, G. (2013). Interview with Grazia Concilio, Città Studi Campus Sostenibile project.

de Jong, M., Joss, S., Schraven, D., Zhan, C. and Weijnen, M. (2015). Sustainable–smart–resilient–low carbon–eco–knowledge cities; making sense of a multitude of concepts promoting sustainable urbanization. *Journal of Cleaner Production*, earlyview online.

Einarsson, F. (2013). Interview with Finnur Friðrik Einarsson, Iceconsult Iceland, MainManager.

ENoLL (2013). *European Network of Living Labs website*. [Online]. Available: www.openlivinglabs.eu [18 September 2015].

Eriksson, M., Niitamo, V.-P., and Kulkki, S. (2005). State-of-the-art in utilizing Living Labs approach to user-centric ICT innovation – a European approach. [Electronic]. Available: http://www.vinnova.se/upload/dokument/Verksamhet/TITA/Stateoftheart_LivingLabs_Eriksson2005.pdf [18 September 2015].

Eskelinen, J., Garcia Robles, A., Lindy, I., Marsh, J. and Muente-Kunigami, A. (2015). *Citizen-Driven Innovation – A Guidebook for City Mayors and Public Administrators*. World Bank and ENoLL. [Electronic]. Available: http://hdl.handle.net/10986/21984 [18 September 2015].

Eurbanlab. (2012). *Eurbanlab Flyer*. [Electronic]. Available: http://www.five.es/descargas/archivos/Eurbanlab_Flyer.pdf [18 September 2015].

European Commission (2013a). *Open and Participative Innovation. Digital Agenda for Europe*. [Online]. Available: http://ec.europa.eu/digital-agenda/en/open-and-participative-innovation [18 September 2015].

European Commission. (2013b). *Open Innovation 2.0 Yearbook 2013*. [Electronic]. Available: https://ec.europa.eu/digital-agenda/en/news/open-innovation-20-yearbook-2013 [18 September 2015].

Evans, J. and Karvonen, A. (2011). Living laboratories for sustainability: exploring the politics and epistemology of urban transition. In Bulkeley, H., Castán Broto, V., Hodson, M. and Marvin, S. (eds), *Cities and Low Carbon Transitions*, Routledge, London. 126–141.

Evans, J. and Karvonen, A. (2014). 'Give me a laboratory and I will lower your carbon footprint!' – Urban laboratories and the pursuit of low carbon futures. *International Journal of Urban and Regional Research*, 38: 413–430.

Farrall, H. (2012). Promoting, innovating and financing urban resilience: a living lab experience. In *Sixth Urban Research and Knowledge Symposium 2012: Research and Policy Papers*. [Electronic]. Available: http://siteresources.worldbank.org/INTURBANDEVELOPMENT/Resources/336387-1369969101352/Farrall-final.pdf [18 September 2015].

Følstad, A. (2008). Living labs for innovation and development of information and communication technology: a literature review. *eJOV: The Electronic Journal for Virtual Organization and Networks*, 10: 99–131.

Garnier, M. (2013). Interview with Mathieu Garnier, 18 July 2013. Urban Living Lab France.

Greenfield, A. (2013). *Against the Smart City (The City Is Here for You to Use Book 1)*. Kindle Edition.

Holst, M. (2013). Interview with Marita Holst, Botnia Living Lab.

Horelli, L., Jarenko, K., Kuoppa, J., Saad-Sulonen, J. and Wallin, S. (2013). *New Approaches to Urban Planning – Insights from Participatory Communities*. Aalto University, Espoo, Finland. [Electronic]. Available: https://aaltodoc.aalto.fi:443/handle/123456789/10244 [18 September 2015].

JPI Urban Europe (2013). *JPI Urban Europe Call for Proposals*. [Online]. Available: www.jpi-urbaneurope.eu/dsresource?objectid=329044&type=org [3 July 2013].

Juujärvi, S. and Pesso, K. (2013). Actor roles in an urban living lab: what can we learn from Suurpelto, Finland? *Technology Innovation Management Review*, 3: 22–27.

Kallio, A., Harmaakorpi, V. and Pihkala, T. (2010). Absorptive capacity and social capital in regional innovation systems: the case of the Lahti region in Finland. *Urban Studies*, 47: 303–319.

Karvonen A., Evans, J. and van Heur, B. (2014). The politics of urban experiments: radical change or business as usual? In Marvin, S. And Hodson, M. (eds), *After Sustainable Cities*. London: Routledge, 105–114.

Kitchin, R. (2014). The real-time city? Big data and smart urbanism. *GeoJournal*, 79: 1–14.

König, A. (2013). Conclusion: a cross-cultural exploration of the co-creation of knowledge in living laboratories for societal transformation across four continents. In König, A. (ed.), *Regenerative Sustainable Development of Universities and Cities: The Role of Living Laboratories*. Cheltenham: Edward Elgar, 273–304.

Krogstie, J (2013). Interview with John Krogstie, Norwegian University of Science and Technology (NTNU).

Krogstie, J., Holst, M., Ståhlbröst, A., Jelle, T., Kulseng, L. and Braskus, L. (2013). Using a living lab methodology for developing energy savings solutions. *19th Americas Conference on Information Systems*. Chicago, 15–17 August 2013.

Leminen, S., Westerlund, M. and Nyström, A.-G. (2012). Living labs as open-innovation networks. *Technology Innovation Management Review*, 2: 6–11.

Liedtke, C., Welfens, M. J., Rohn, H. and Nordmann, J. (2012). LIVING LAB: user-driven innovation for sustainability. *International Journal of Sustainability in Higher Education*, 13: 106–118.

Loorbach, D. (2013). Interview with Derk Loorbach, Dutch Research Institute for Transitions (DRIFT).

Luque-Ayala, A. and Marvin, S. (2015). Developing a critical understanding of smart urbanism? *Urban Studies*, 52: 2105–2116.

McCormick, K., Anderberg, S., Coenen, L. and Neij, L. (2013). Advancing sustainable urban transformation. *Journal of Cleaner Production*, 50: 1–11.

Markopoulos, P. and Rauterberg, G.W.M. (2001). *LivingLab: A White Paper, 35.* [Online]. Available: www.idemployee.id.tue.nl/p.markopoulos/downloadablePapers/LivingLabWhitePaper.pdf [18 September 2015].

Melkas, H. and Harmaakorpi, V. (2008). Data, information and knowledge in regional innovation networks. *European Journal of Innovation Management*, 11: 103–124.

Mensink, W., Birrer, F. A. and Dutilleul, B. (2010). Unpacking European living labs: analysing innovation's social dimensions. *Central European Journal of Public Policy*, 1: 60–85.

MIT News (2003) Lab homes in on home life. [Online]. Available: http://news.mit.edu/2003/lab [18 September 2015].

Moor, K.D., Ketyko, I., Joseph, W., Deryckere, T., Marez, L.D., Martens, L. and Verleye, G. (2010). Proposed framework for evaluating quality of experience in a mobile, testbed-oriented living lab setting. *Mobile Networks and Applications*, 15(3): 378–391.

Nevens, F., Frantzeskaki, N., Gorissen, L. and Loorbach, D. (2013). Urban transition labs: co-creating transformative action for sustainable cities. *Journal of Cleaner Production*, 50: 111–122.

Pallot, M., Trousse, B., Senach, B. and Scapin, D. (2010) *Living Lab Research Landscape: From User Centred Design and User Experience towards User Co-creation.* First European Summer School 'Living Labs', August 2010, Paris. Available: https://hal.inria.fr/inria-00612632/document [1 February 2016].

Pallot, M., Krawczyk, P. and Kivilehto, A. (2013). User centred open innovation domain landscape within the European Network of Living Labs. Paper presented at *Challenges for Sustainable Growth in Helsinki, Finland*, 16–19 June 2013.

Park, R.E. (1915). The city: suggestions for the investigation of human behavior in the city environment. *American Journal of Sociology*, 20: 577–612.

Pascu, C. and van Lieshout, M. (2009). User-led, citizen innovation at the interface of services. *Info*, 11: 82–96.

Peng, M.W. (2010). *Research Streams*. University of Texas at Dallas. [Electronic]. Available: http://www.utdallas.edu/~mikepeng/documents/CV201101_ResearchStream.pdf [18 September 2015].

Periphèria. (2011). *Periphèria Project – Networked Smart Peripheral Cities for Sustainable Lifstyles*. [Online]. Available: http://www.peripheria.eu [3 July 2013] and http://humansmartcities.eu/project/peripheria [18 September 2015].

Schneidewind, U. and Scheck, H. (2013). Die Stadt als „Reallabor" für Systeminnovationen. In Rückert-John J. (ed.), *Soziale Innovation und Nachhaltigkeit*. Wiesbaden: Springer Fachmedien, 229–248.

Schliwa, G. (2013). *Exploring Living Labs through Transition Management - Challenges and Opportunities for Sustainable Urban Transitions.* Master's Thesis, International

Institute for Industrial Environmental Economics, Lund University. [Online]. Available: http://www.lunduniversity.lu.se/lup/publication/4091934 [15 September 2015].

Schuurman, D., De Moor, K., De Marez, L. and Evens, T. (2011). A living lab research approach for mobile TV. *Telematics and Informatics*, 28: 271–282.

Schuurman, D., Lievens, B., De Marez, L. and Ballon, P. (2012). Innovation from user experience in Living Labs: revisiting the 'innovation factory'-concept with a panel-based and user-centered approach. In *Conference Action for Innovation : Innovating from Experience, Proceedings*. Ghent: Ghent University, Department of Communication Studies.

Schuurman, D. (2015). Living labs: a systematic literature review. Paper presented at *ENoLL OpenLivingDays 2015*. [Online]. Available: http://www.scribd.com/doc/276089123/ENoLL-Research-Day-Conference-Proceedings-2015#scribd [15 September 2015].

Sherwood, H. (2014). How to create happy communities through co-housing. *The Guardian*, 21 November 2014. [Electronic]. Available: http://www.theguardian.com/cities/2014/nov/21/how-to-create-happy-communities-through-co-housing [24 February 2015].

Ståhlbröst, A. (2012). A set of key principles to assess the impact of living labs. *International Journal of Product Development*, 17: 60–75.

Ståhlbröst, A. (2013). Interview with Anna Ståhlbröst, Botnia Living Lab.

Ståhlbröst, A. and Holst, M. (2013). *The Living Lab Methodology Handbook*. [Online]. Available: http://www.openlivinglabs.eu/news/living-lab-methodology-handbook [18 September 2015].

SusLabNWE. (2012). *SusLabNWE Partnerbroschüre*. [Online]. Available: suslab.eu/fileadmin/suslab/BrochurePrint1.pdf [18 September 2015].

Talwar, S., Wiek, A. and Robinson, J. (2011). User engagement in sustainability research. *Science and Public Policy*, 38: 379–390.

ULL-VSQY (2013). *Homepage Urban Living Lab VSQY*. [Online]. Available: http://www.urbanll.com [July 2013].

Viitanen, J. and Kingston, R. (2014). Smart cities and green growth: outsourcing democratic and environmental resilience to the global technology sector. *Environment and Planning A*, 46: 803–819.

von Hippel, E. (1986). Lead users: a source of novel product concepts. *Management Science*, 32: 791–805.

Voytenko, Y., McCormick, K., Evans, J. and Schliwa, G. (2015). Exploring urban living labs for sustainability and low carbon cities in Europe. *Journal of Cleaner Production*, earlyview online.

Wallin, S. (2014). *APRILab: Guidelines to Define and Establish an Urban Living Lab*. University of Amsterdam, Amsterdam Institute of Social Science Research. [Online]. Available: http://aissr.uva.nl/research/externally-funded-projects/sites/content13/aprilab/deliverables/deliverables.html [18 September 2015].

World Bank. (2012). *World Bank, ENoLL Strengthen Cooperation on Open Innovation*. [Online]. Available: http://www.worldbank.org/en/news/feature/2012/07/13/world-bank-enoll-strengthen-cooperation-on-open-innovation [18 September 2015].

Yin, R. K. (2003). *Case Study Research: Design and Methods*. Thousand Oaks, CA: Sage.

Yvelines Conseil Général (2010). *Un Projet d'Expérimentation des Véhicules Electriques dans les Yvelines*. [Online]. Available: https://www.yvelines.fr/2010/04/13/un-projet-dexperimentation-des-vehicules-electriques-dans-les-yvelines [18 September 2015].

Part III

Experimental cities

13 Turning over a new leaf

Sustainability and urban experimentation in Seoul

Sofia T. Shwayri

Introduction

South Korea's recent transformation from a poor and war-ravaged nation in the 1950s to a global economic power of the twenty-first century has been achieved, in large part, by a particular Korean blend of philosophy, hard work and adapting to serious external pressures of the financial, military and environmental kind. As there was no precedent in both scale and achievement, the approach was largely experimental. Today, the established industrial nation is applying that uniquely Korean experimental approach to achieve sustainability in the new millennium. South Korea's capital city, Seoul, owes much of its present-day urban form to policies and developments of the 1960s, specifically the second Korean Five-Year Plan that covered the years 1966 to 1970 when demand for housing and services rose exponentially in line with rapidly expanding per capita income. During those years, South Korea was ruled by a military dictatorship focused on national development based on export-led economic growth. These policies were intended to bring the nation to the upper echelon of industrialized nations as fast as possible. Starting as one of the poorest nations in 1960, the Korean economy grew at an average rate of 9.6 per cent between 1967 and 1972. The Seoul Municipal Government (SMG) during the years of this plan was headed by Kim Hyun Ok (1966–1970), a former military man and confidant of President Park Chung Hee (1961–1979). He earned the nickname 'Bulldozer' owing to his enthusiastic and hands-on execution of mega developments that included housing, retail and commercial buildings, widening of the Han River and construction of new roads and overpasses (Hong 2013). Growth continued throughout the 1970s and into the1980s with Seoul receiving at least its fair share of the rewards. In 1978, SMG bid to host the 1988 Summer Olympic Games, winning that privilege in 1981 (Kang 2004). In preparation for this mega event the capital went through a major upgrade to its infrastructure and image, putting Seoul on the map of world cities, posing a permanent challenge to its future leaders to maintain this newly acquired status. In subsequent years, efforts have focused on competition with existing regional and global cities through large-scale projects within the city boundaries, such as the Han River Redevelopment Project that included the Magok district and Sangnam New Millennium Town, with the later emergence of Digital Media City.

The 1990s, however, proved more problematic. The SMG faced the challenge of dealing with a crumbling infrastructure hastily established in the era of rapid growth, while the central government was buffeted by the fallout from the 1994 nuclear crisis and the 1997 Asian financial crisis. Most large-scale projects were stalled, then repurposed and re-launched as the new millennium got under way, when it became ever more urgent to prioritize remedies for failing infrastructure that included major public landmarks such as department stores, bridges and highway overpasses in Seoul (Kong 2014). The 1997 financial crisis in particular forced the central government to restructure the national economy, making it more service-oriented and encompassing a new mode of urban development centred on ubiquitous technologies and foreign direct investment. Since becoming an OECD member in 1996, the government has increasingly been asserting its role as an active member of the global community by taking a number of key steps, including the ratification of the Kyoto Protocol in November 2002. The real work, however, began later in the decade following President Lee Myung-bak's speech at the G8 meeting in Tokyo in 2008. He promised that Korea's commitment was not only going to be national but global, to be achieved by driving the expansion of the East Asian Partnership into one of Global Green Growth, so signalling a shift in its level of commitment towards reducing carbon emissions by serving as the mediator between developing and developed nations. As a measure of the president's commitment to climate change issues, Lee Myung-bak used the occasion of the nation's sixtieth anniversary in 2008 to announce the Green Growth policy as a new paradigm for development. A white book titled 'Green Growth of a Greater Korea' was published detailing both policy directions and its core projects (Kim and Choi 2013). Some of the stalled projects became attractive again as they contained elements beneficial for the promotion of climate change initiatives within Korea and as model projects for export, placing the nation's growing expertise in this area firmly in the international spotlight.

For the SMG, Green Growth translates from national policy to local action with the creation of a more liveable and healthier city as a way to improve the quality of life for its citizens by repurposing existing developments and upgrading and redesigning failing infrastructure. While the Green Growth paradigm emphasizes the development of new engines of economic growth, it is the societal transformation that it embodies which is fundamental to attaining a green society that is highly efficient in low carbon energy (one of three key goals outlined by the president) as well as making Korea a world leader in this new mode of development (Kim and Choi 2013). The other two goals are the development of an industry-leading green economy and the creation of Korea as a world leader in Green Growth (Kim and Choi 2013). Korea could only achieve this status if it improved the quality of life of its people and consequently improved the nation's reputation amongst the international community. New partnerships between government, the private sector and the people were to be created by pushing for more 'active participation and cooperation' from those groups (Kim and Choi 2013: 14). For the SMG, this meant becoming more transparent by fighting corruption and inviting Seoulites to get involved in urban issues through accessing information and sharing opinion. These changes would not only strengthen e-government but contribute to the emergence of a smart city.

Driven fundamentally by climate change concerns, these practices are ushering in a new era of urban governance very much in sync with the global phenomenon of urban experimentation (Bulkeley and Castán Broto 2013). The adoption of a Green Growth vision acknowledges the challenges posed by climate change and energy consumption. It attempts, through an integrated social approach of collaboration amongst governments, private enterprise and civil society, to identify solutions to the myriad issues faced by communities big and small. This integration allows for 'deep understanding, swift decision-making, revolutionary innovations, and empathetic approaches' to embrace a set of strategies translated into policies and executed through a series of projects (Park 2013). Opportunities arose that lent support to developments that had been mired by controversies, as well as providing the means for proposing new ones. Climate change experiments do not have to be designed with just this purpose in mind but, because of the ubiquity of this discourse (Bulkeley and Castán Broto 2013), climate change might become one of the goals and thus it is important to understand how projects are conceptualized by examining their specific circumstances. Urban experiments are revolutionary innovations that engender new leadership, new forms of partnerships, new forms of knowledge production and consequently new practices, with some pointing to new types of spaces in the experimental city (Evans and Karvonen 2011; Castán Broto and Bulkeley 2013; Kullman 2013).

In an experimental city, change is built into the planning process (Spilhaus 1967; Karvonen *et al.* 2014), as is fast adoption and adaptation (Karvonen and van Heur 2014). In Seoul (and Korea in general), change is built into the psyche of government, businesses and people. 'Bali bali' (hurry hurry) is the mantra that has driven Koreans throughout the years of phenomenal post-war growth both in terms of rapid adaption and high-speed development (Forsyth 2012). The sense of catching-up that has characterized Korea's process of modernization has long shaped Seoul's development since the 1960s. Operating in compressed time, adopting and adapting foreign ideas and policies at the speed of light, and sometimes reversing them at the same speed in favour of the latest prominent concept, summarizes planning practice. Fast adaptation has created a 'compressed modernity' moving the Korean nation in one generation from a backward and poor society to one of the most technologically advanced on the planet (Chang 2010), becoming the envy of many governments in the developing world and seen as a model to imitate. The new towns adopted and adapted during this period of modernization assumed a distinctly Korean flavour and have since become a model form of development that has been exported to countries in Southeast Asia and the Middle East. In this latest round of development, cities, especially Seoul, are being turned into living laboratories for addressing climate change challenges as the means to maintain competitiveness while assuming a leadership position both regionally and globally.

In any laboratory, there can be many experiments taking place both in parallel and sequentially, some more mundane than others, some requiring huge resources and others few, and some providing results that look familiar to others but that have been reached by completely different processes. In this respect, Seoul as an urban laboratory is no different than any vibrant city whose government is facing both its immediate and distant challenges in a way that maintains its economic competitiveness. The SMG (2014) four-year plan introduced in September 2014 envisions a 'people centered Seoul, citizen happy Seoul' based on creating a safe and economically vibrant metropolis. The plan has been translated, through the integrated social approach, into a number of initiatives and projects where scale determines the degree of participation between citizens, the market and the public sector. Although very different in scale, two of these experiments are equally transformative for the capital in terms of spatial impact, economic growth and the reduction of its carbon footprint. These projects, Magok eco-town in the west of the capital, the 'last undeveloped site in Seoul' (SMG 2012: 4), and the Seoul Overpass Park (the 7017 Project, the naming an amalgam of the build year, 1970 and its height, 17 metres) (SMG 2015) near Seoul Station in downtown Seoul (Figure 13.1) are the subjects of this chapter. They are not meant to be treated comparatively but rather are used to illustrate that they have similar economic impacts despite being very different in terms of scale and the partnerships engaged – the former mainly one of private enterprise and the latter driven by the public sector. The development of Magok will deliver a resource-hungry Northeast Asian regional Research and Development district, while the downtown project will see a relatively low-cost, High Line-style elevated park rising from the rust of a failing 1970s-era inner city overpass, a key component to realizing a walkable city vision making Seoul even more attractive as a tourist node (Shwayri forthcoming).

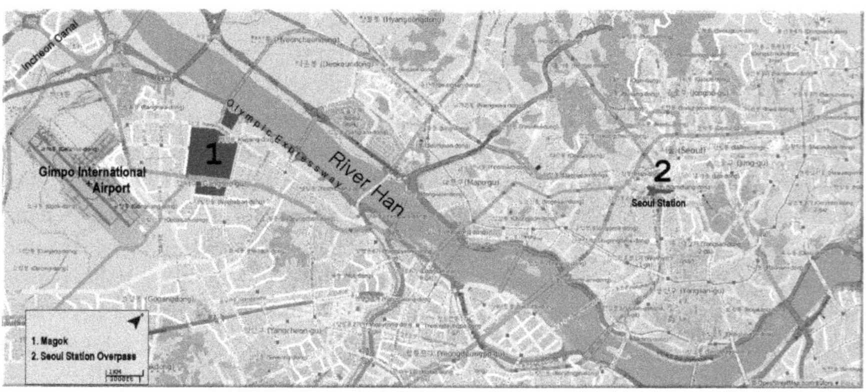

Figure 13.1 Western and Central Seoul showing Magok (1) and Seoul Station (2) (source: adapted from OpenStreetMap by David Anderson).

After Songdo: Magok, the development of Korea's second 'Northeast Asian Hub'

'Sitting quietly does not lead to a miracle,' said Korean president Park Geun-hye in Seoul on 23 October 2014 at the groundbreaking ceremony for LG's proposed Science Park (Limb 2014). It was part of a speech in praise of the chaebol's (family-run conglomerate) commitment to a future of challenge in the fiercely competitive world of technology and telecommunications. She could well have been talking about the district in the capital – Magok – on which this new development is situated and for which a role has finally been identified. Construction plans were announced in December 2005 for the Magok District but there was a two-year wait before an area was designated for the development and a further two years before construction would commence. Thus began the 'Eco Energy Town of the Future' with an emphasis on energy efficiency. Heating and air-conditioning systems would need half as much fuel and energy with 40 per cent of the area's energy being recycled. New buildings would require only two-thirds of the energy used by the average building while 10MW of energy needs would be met by solar power, thus contributing to the city's reduction in greenhouse gases. Winter heating for 23,000 houses would be provided by energy derived from the River Han. And finally, the R&D centre in the development would explore further energy efficiencies in the area (The Korea Times 2008). Soon, however, the SMG decided to rebrand the Magok development, referring to it as a regional hub in the middle of Seoul or the 'Knowledge Industrial Green City Leading the Future of Northeast Asia'. The goal was to create a gateway city to Northeast Asia that was simultaneously an innovative base of the knowledge industry and a green city of the future. It would be open to everyone; a convergence site of cutting-edge technology and industry, and would aim for sustainable value (SMG 2012).

In form and function, the Magok development contrasts starkly with the Songdo International Business District (IBD), part of the Incheon Free Economic Zone (IFEZ), the northeast Korean business utopia. Figure 13.2 shows the contrasting cityscapes of the two projects. Magok is situated south of the River Han in western Seoul in Gangseo-Gu and east of the Incheon Canal, the Airport Expressway and Banghwa-dong. To its southwest, less than three kilometres away and linked directly by road is Gimpo Airport (putting the district about two hours flying time away from major cities in China and Japan). Deungchon-dong lies to the east. Three rail lines pass through it (the Airport Express line that connects Seoul to Incheon International Airport and Seoul Metro lines 5 and 9). Separating it from the River Han is the Olympic Expressway to the north. Its location at the extreme western end of the city is also marked by the start of a more militarized area as the River Han, with barbed wire and watchtowers along its banks, flows northwest towards its mouth, close to the Northern Limit Line, the disputed maritime border in the Yellow Sea separating South and North Korea. Any maritime traffic that plans to use the waterway here has to obtain UN permission. The Incheon Canal acknowledges this fact, allowing shipping to take a safer and more direct path to the coastal port of Incheon and beyond.

Figure 13.2 Songdo (left), the Compact Smart City, and Magok (right) (source: David Anderson).

Like Magok, Songdo IBD, 40 miles southwest of Seoul, is also under development but at a more advanced stage. It is situated on reclaimed land in Incheon on the northwestern coast of the Korean Peninsula, also in close proximity to the Northern Limit Line. A 15-minute trip over the newly built Incheon Bridge separates Songdo from Incheon International Airport, making it less than three hours from a quarter of the world's population and two of the largest economies, Japan and China. Its location and connectivity has been celebrated by some as 'an experimental prototype of the aerotropolis' (DiNardo 2013). As a high-tech or ubiquitous city, Songdo is planned around major industries (including biotechnologies) as well as being a hub for services, particularly educational (with the establishment of global campuses) and medical (through its international hospital) services.

Both developments have a longer history. Songdo goes back to 1988 while Magok has undergone more than ten iterations of planning from 1995. It was touted in 2007 by then Mayor Oh Se-hoon as part of a multi-year project that was to turn Seoul into 'an attractive waterfront city with high-level tourist and transportation facilities' (The Hankyoreh 2007). This Han River Renaissance project, consisting of 33 sub-plans including one for Magok, eventually foundered on the rocks of the 2008 financial crisis as well as resistance from environmentalists and serious project delays (The Hankyoreh 2007). Previously, Magok had been earmarked for several projects that were initiated, punctuated or halted by outside events writ large in the history of Korea – financial crises, inter-Korean crises, regional mega projects and international sporting events. In the decades leading up to its current incarnation, the district was, among other things, a candidate site for Seoul's World Cup stadium which was ultimately sited north of the Han River.

The influences on the Magok development are reflected in the built environment, especially when contrasting the new city with its more famous predecessor, Songdo. In terms of land, Magok required no reclamation, no waterway connection but there is a central park feature, an essential requirement made by the city in the International Design Competition. This is a Korean regional hub, as opposed to being global or international, so there is no need for enticements for foreign investment such as international university campuses, memoranda of understanding with international concerns or technology contracts with overseas corporations. The communication infrastructure already exists due to previously planned projects (e.g. subway stations on Seoul Metro Lines 5 and 9 as well as the Airport Express). Access to air transport is provided by Gimpo Airport, which is a regional rather than international airport like Incheon. Finally, housing reflects the Korean reality of a smaller family unit rather than the lavish international penthouse styles of Songdo's apartment structures. Songdo IBD's global aspirations were emphasized by aggressive media campaigns, mayoral tours of the United States and other high profile activities designed to attract foreign investment (Kim 2010, Shwayri 2013). Magok's focus has always been regional while the SMG's opportunity to attract international attention has not been overlooked. One such example was the presentation of the Magok development plans by the Mayor of Seoul at a pre-C40 summit meeting in May 2009 to an audience that included Bill Clinton, a key player in the Cities Climate Leadership Group. This led to an MOU (Memorandum of Understanding) between Seoul City and the Clinton Foundation to implement a 'climate positive development program' (Clinton Foundation 2009). Magok was chosen as the site for the signing (Kwon 2009).

Magok's time has come. It is the last sizable area of land for development within the Seoul Metropolitan Area. Its previous competitors, including Mokpo, Jamsil, Yongsan, Yeoido and Gangnam are not immediately available for further development. In its new role, Magok benefits largely from lessons learned in other projects of a similar scale across Korea. These include the provision of housing that people need (as opposed to what urban plans provided), recognition of the nation's profitable or growth markets (emphasizing Korea's regional leadership) and the associated transportation hubs that connect these markets. Seoul, much like the rest of Korea, has experienced population decline and a growing demand for housing for singles or childless couples rather than the government supplied housing that continues to cater to nuclear families (Pressian 2004). Housing under development in Magok recognizes this reality with its planned supply of one- and two-bedroom apartments. In terms of markets and competition, these are much closer to home and more focused on Korean expertise than the hoped-for global playing fields catered to by developments such as Songdo out to the west. A highly educated workforce is, in reality, going to be predominantly Korean, with foreigners playing a role much below those in the reach and scope planned in Songdo and the other Free Enterprise Zones. Magok's scale is also cognizant of a new reality, touted by the current Mayor of Seoul as the 'paradigm shift' in urban development from the megalomaniacal new build projects to the more sustainable developments appreciative of Seoul's traditions (SMG 2014).

Sitting quietly may not have led to a miracle but in Magok's case, it has produced reality, even after many attempts of experimenting with different forms and functions (Seoul Institute 2014). The lessons learned from previous experiments both on this site and elsewhere were put to good use in the Magok of today.

The district was part of the ill-fated River Han Renaissance Project (2007) as well as being a touted site for the Seoul 2002 World Cup Stadium (1996). Experiments elsewhere also provided valuable input. For example, Songdo IBD never got its expected foreign contingent and only thrived when housing was opened up to Koreans after 2010. Turning infrastructural failure into a positive outcome while at the same time enhancing the nation's competitive edge in tourism is the underlying theme of a distinctly different experiment taking place in downtown Seoul – the Seoul Overpass Park.

A park in the sky: repurposing of failed infrastructure

The Seoul Overpass Park, an elevated recreational area born out of defunct post-Korean War infrastructure, will soon join its older sibling, the hugely successful Cheonggyecheon Park, in downtown Seoul. Both are products of nearly identical circumstances. They have been promoted in similar fashion politically and possess personas more modern than their admiring public believes. The literature concerning the latter is long and voluminous (Cho 2010; Lee and Anderson 2013). Discussion about the former, however, mainly concerns its outward similarity in form, function and history to New York City's world famous High Line Park (Jung 2014; Kwaak 2014).

Loved during their heyday, promoted with God-like reverence as beacons of modernity and implemented gleefully by civic leaders more reminiscent of medieval warrior kings (Mesmer 2014), Seoul's network of overpasses criss-crossed the city. They were implemented at a fraction of the cost and construction time of extending the subway system. They funnelled workers from the rapidly growing suburbs to their workplaces downtown. Within a generation, however, decay had set in and the overpasses required extensive repairs with maintenance costs increasing at an astronomic rate (Koh 2013). Additional safety measures saw mandated decreases in traffic flows of up to 50 per cent. The story of the Ahyeon Overpass is one example of more than 100 overpasses built at much the same time and that were, by the late 1990s and early 2000s, placing a tremendous repair and maintenance burden on the coffers of the SMG. The overpass at Ahyeon was the first to be built. One kilometre in length, erected in 1968 and planned to move 80,000 vehicles per day between central Seoul and Mapo, Chungjeongno and Sinchon, it had a lifespan of 36 years. In 2004, less than a decade after the Sampoong Department store collapse had begun to sow doubts about the safety of Seoul's crumbling infrastructure (World History Project 1995), US$7.19 million was spent on repairs and the weight capacity of the rotting structure was halved. By this time, maintenance costs were running at an annual rate of US$25 million. The Ahyeon overpass limped on for a further ten years until it was finally demolished in 2014 (McKeag 2014). Overpasses also suppressed commercial and residential development in a capital city becoming denser – not just around the Central Business District but elsewhere. The dangers their loads posed to air quality were well documented. Locals also reported that the spaces under overpasses were often used for illegal parking or for unloading construction materials.

Residents were averse to walking below them with little sunshine and few decent vistas (Koh 2013). Overpasses were no longer a sustainable feature of a modern, progressive metropolis. Furthermore, these older structures were physical barriers to growth in areas such as the capital's Central Business District, forcing lucrative commercial concerns, together with people and housing, out to suburbs like Gangnam. The environmental impacts of the rapid growth era, if known or fully understood at the time, featured very low on the government's list of priorities.

More famous is the story of the Cheong Gye overpass that was removed in 2002 and the project that replaced it: a four-kilometre-long inner city park whose central feature is the re-exposed Cheong Gye stream, significant in the history of Seoul but buried in the name of progress. Contrary to widespread belief, what has been 'uncovered' is not the original stream but one following the course of its predecessor, fed by very modern pumping stations and serving as a major component in Seoul's flood defences (Mesmer 2014). The environmental advantages it displays over the structure it replaced, while hugely relevant, are by-products of the original economic and financial benefits touted for its construction (Lee and Anderson 2013). The Seoul mayoral campaign of 2002, for which this was a key election issue, highlighted the project bringing economic vitality to the inner city. It was also a device by which the incoming mayor and future president, Lee Myung Bak, could expand his political influence into northern Seoul, where it was weak at the time (Cho 2010). In a similar manner, the stream's banks and restraining walls are largely contemporary in both materials and design, serving to open up access to the neighbouring businesses, as high-profile stages for the promotion of those concerns, as well as the more mundane functions served by a public park. In these respects, Cheonggyecheon (and as we will see later, the Seoul Overpass Park) is similar in nature to the Seoul Fortress walls – reconstructions of an earlier reality.

Today, Seoul city officials recognize overpasses as relics of a bygone era that now restrict urban planning for the twenty-first century. Seoul Overpass Park, as well as the decade-older Cheonggyecheon, help mitigate the expensive solutions to problems of decay in the infrastructure that underpinned Seoul's rapid growth from the late 1960s through the 1970s. Around 16 overpasses have been torn down since 2002 with just 84 of the original number remaining. The cleared sites present opportunities for urban experiments that include dedicated bus median lanes, expansion of available development space, green spaces and tourism, as well as providing ecologically friendly areas. These benefits were quickly picked up by politicians with local, national and worldwide interests, and packaged, among other ways, as sources for improvements to the quality of life for Seoul's citizens through the provision of 'views and beauty of the city'. Soon after the Ahyeon overpass was finally removed, the SMG announced plans to develop 'twenty-five strategic tourist destinations that can present their historic resources and dynamism to tourists, as core parts of a city that has served as the nation's capital for 2,000 years' (SMG 2013). Seoul, with a relatively static population today of 10 million, sees growth in tourism as a major source of city revenue. Over 12 million people visited the city in 2013, up from 7 million just three years earlier, making the Korean capital one of the most visited cities in the world (Lee 2014). Cheonggyecheon alone receives around 20 per cent of these visitors. In conjunction with the increase in real-estate values around the stream, the local economy has received a significant boost (City Clock Magazine 2014).

To maintain and encourage further tourism, in 2014 the SMG announced a New Urban Architecture Road Map for 2014–2018, with the slogan 'Walking City', aiming to reduce the number of roads and improve pedestrian environments (Yoon 2014). This new road map includes 25 core policies focusing on cutting back roadways to encourage pedestrian traffic, setting up the Urban Restoration headquarters and designing, among other projects, Gaepo Digital Innovation Park, Hongneung Smart Aging Cluster, Seoul Overpass Park and Magok Central Plaza. Roads around the four main gates of Seoul will be gradually removed for the benefit of pedestrian access and to create a public park, and bicycles are favoured over cars. The major urban restoration projects in the plan include the preservation of the Seun Sangga and the remodelling of the vehicle-only Seoul Station overpass to create a linear urban park (Figure 13.3).

Since the landslide re-election of Seoul Mayor Park Won Soon in June 2014, policy is now focused on the process of restoring original form and preservation. The mayor stated: 'the paradigm of "good development" has shifted [from a penchant for the modern to respect of the past, and] the [Seoul Station overpass] project will be a good example of the paradigm change'. Furthermore 'it is better to rejuvenate the city instead of destroying its cultural heritage to build something new' (Jung 2014). In its safety inspection in 2008, Seoul Station overpass received

Figure 13.3 The old Seoul Station (left) and its overpass (source: David Anderson).

a level D (only level E is worse). This will be transformed into an ecological citizen culture park. In fact, this experiment has already been prototyped. On 4 October 2014 the overpass was closed to traffic for four hours and the public were invited to stroll around the elevated space from access points in the commercial and tourist centric district of Namdaemun. This was repeated with a further four-hour pedestrian-only opening on 10 May 2015. In October 2014, an international design competition was held and construction will begin in 2015. It will preserve the original form but will largely be a reconstruction, introducing new content by replacing plates and repairing other weakened components in the name of safety.

Seoul's Overpass Park was announced in the Korean press on 1 September 2014. At Mayor Park Won-soon's suggestion, the 17-metre-high overpass will be closed to motorized vehicles and renovated as a park for pedestrians at an expected cost of US$37.5 million to be funded by the SMG (Jung 2014). Seoul was said to have been inspired by the High Line, the world famous urban-renewal project in New York's Lower West Side. A boon for tourism as well as real-estate values in the area, the one-mile-long linear park was opened in 2009 on a disused elevated rail line, itself modelled after the Promenade Plantée in Paris. While in New York for the UN Climate Summit on the 23 September 2014 the mayor announced the Seoul Overpass Park project to the world. Inspiration may very well have come from New York but the drivers, motivation and opportunity for this and similar projects arose from the legacy of decay in the infrastructure of a city rapidly built to climb onto the world stage and which now needs to maintain that elevated position through reducing maintenance overheads, generating income from tourism and the ongoing support of the city's voters who require perceived improvement in the quality of their urban lives.

Conclusions

Seoul is not a demonstration or model city attempting to show what can be done temporarily to renew itself. Seoul is an urban laboratory, an experimental city. Some experiments may be undertaken that signal a sense of temporality, as they are reversed if their anticipatory purpose falls short of expectations, such as projects that address a particular aspect of climate change. These may look like simple interventions using novel ideas or practices for sustainable modes of urbanism. When successful, these urban experiments and their manifestations can bring about lasting transformations and act as models to be replicated. Fast change engendering such forms of urban development appears to be novel and attractive in its potential for innovation. In Korea, however, change and the consequent sense of temporality has been part and parcel of urban development since the 1960s, as well as the city being a model of urban development for other aspiring nations. In less than one generation, Seoul moved from an impoverished capital at the periphery of the world economy to its centre; a process commonly referred to as 'compressed modernization'. Inherent in this process is change, constant adoption and adaptation of the foreign that is new, fast, superfast or à la Korean bali bali. This is ever more important for Seoul to maintain its economic competitiveness in the race amongst cities. Economic considerations have long triggered this change and hence the sense of temporality embedded in the landscape fostering new modes of urban developments.

In this 'Age of Crisis' (Caprotti 2015), the failure of addressing the global climate challenge for over two decades has belatedly triggered much of the urban climate experimentation, often hailed as temporary until an assessment of the outcomes can turn them into more permanent solutions. It is both failure and temporality that underlies much of the current developments in Seoul and is the foundation for more permanent transformations. Failed developments present opportunities to repurpose outdated infrastructure in experimental ways that accrue economic benefits and help government and media sell these to the local and global public as climate initiatives. Seoul is a city whose image of modernity was built on this intricate web of modern infrastructure and one that champions Seoulites and their well-being without compromising economic benefits. The Seoul Overpass Park will create a climate-friendly area with the potential to reinvigorate the walkable city that the core once was while serving the tourist sector. Similarly, it is economics that saw Magok turn from an 'eco-energy town of the future' to a 'Knowledge Industrial Green City' in a very short period, not because the latter addresses climate change challenge in a better way, but to serve diverse economic sectors. Urban leaders, fully aware of the potential benefits, have made the economy part and parcel of their urban development approach, always ready to experiment with new ideas, sometimes even before they have been put into practice anywhere else, as in the case of eco-energy. Seoul has always adapted lessons learned from previous urban experiments. For example, many of today's environmentally friendly traffic management systems arose from the need to deal with deadly collapses of infrastructure that occurred decades before their modified implementation, while the Bus Rapid Transit lanes that run down the centre of roads are a throwback to the original tram system model removed in the mid-1960s. In driving to remain in the lead group of world cities, Seoul's leadership may even abandon the current 'latest' trends in favour of even more up-to-date versions. These approaches result in processes that can only be achieved by treating Seoul as the ultimate urban laboratory that it is.

References

Bulkeley, H. and Castán Broto, V. (2013). Government by experiment? Global cities and the governing of climate change. *Transactions of the Institute of British Geographers*, 38: 361–375.

Caprotti, F. (2015). *Eco-cities and the Transition to Carbon Economies*. Basingstoke: Palgrave Macmillan.

Castán Broto, V. and Bulkeley, H. (2013). A survey of urban climate change experiments in 100 cities. *Global Environmental Change*, 23: 92–102.

Chang, K. (2010). *South Korea under Compressed Modernity: Familial Political Economy in Transition*. London: Routledge.

Cho, M. (2010). The politics of urban nature restoration: the case of Cheonggyecheon restoration in Seoul, Korea. *International Development Planning Review*, 32: 145–165.

City Clock Magazine (2014). Removing urban highways: the story of the Cheonggyecheon Stream in Seoul. *City Clock Magazine: Thoughts on Cities*. [Electronic]. Available: http://www.cityclock.org/removing-urban-highways/#.VM6J8C4QHpy [20 November 2014].

Clinton Foundation (2009). Press Release: Clinton climate initiative to demonstrate model for sustainable urban growth with projects in 10 countries on 6 continents. Clinton Foundation, 18 May 2009. [Online]. Available: https://www.clintonfoundation.org/main/news-and-media/press-releases-and-statements/press-release-clinton-climate-initiative-to-demostrate-model-for-sustainable-urb.html [10 October 2014].

DiNardo, K. (2013). Songdo, South Korea: the city that could change the way we travel. *The Washington Post*, 3 January 2013. [Electronic]. Available: http://www.washington-post.com/lifestyle/travel/songdo-south-korea-the-city-that-could-change-the-way-we-travel/2013/01/03/babb96a0-4614-11e2-8061-253bccfc7532_story.html [30 May 2014].

Evans, J. and Karvonen, A. (2011). Living laboratories for sustainability: exploring the politics and epistemology of urban transition. In Bulkeley, H., Castán Broto, V., Hodson, M. and Marvin, S. (eds), *Cities and Low Carbon Transitions*. London: Routledge, 126–141.

Forsyth, L. (2012). Under pressure: Byun Ho San-Part 1. *LucForsyth*. [Online]. Available: http://lucforsyth.com/2012/01/under-pressure-byun-ho-san-part-1/ [1 February 2014].

Hong, S. (2013). Seoul, a Korean capital. In Perera, N. and Tang, W. (eds), *Transforming Asian Cities: Intellectual Impasse, Asianizing Space, and Emerging Translocalities*. London: Routledge, 20–32.

Jung, M. (2014). Seoul to have its own 'High Line Park'. *The Korea Times*, 25 September 2014. [Electronic]. Available: http://www.koreatimesus.com/seoul-to-have-its-own-high-line-park/ [10 November 2014].

Kang, H. (2004). Mega events as urban transformer: the experience of Seoul. *Seoul City Research*, 5: 1–15.

Karvonen, A. and van Heur, B. (2014). Urban laboratories: experiments in reworking cities. *International Journal of Urban and Regional Research*, 38: 379–92.

Karvonen, A., Evans, J. and van Heur, B. (2014). The politics of urban experiments: radical change or business as usual? In Hodson, M. and Marvin, S. (eds), *After Sustainable Cities?* London: Routledge, 104–115.

Kim, C. (2010). Place promotion and symbolic characterization of New Songdo City, South Korea. *Cities*, 27: 13–19.

Kim, S. and Choi, H. (2013). *Green Growth for a Greater Korea: White Book on Korean Green Growth Policy, 2008–2012*. Seoul: Korea Environment Institute.

Koh. M. (2013). Removal of overpasses. *The Dong-A Ilbo*, 16 March 2013. [Electronic]. Available: http://english.donga.com/srv/service.php3?biid=2013031634108 [24 November 2014].

Kong, K. (2014). South Korea's history of building collapses. *The Wall Street Journal*, 18 February 2014. [Electronic]. Available: http://blogs.wsj.com/korearealtime/2014/02/18/south-koreas-history-of-building-collapses/ [30 January 2015].

Kullman, K. (2013). Geographies of experiment/experimental geographies: a rough guide. *Geography Compass*, 7: 879–894.

Kwaak, J. (2014). Seoul plans high line style elevated park. *The Wall Street Journal*, 1 September 2014. [Electronic]. Available: http://blogs.wsj.com/korearealtime/2014/09/01/seoul-plans-high-line-style-elevated-park/ [10 November 2014].

Kwon, M. (2009). Clinton calls for climate action for grandchildren. *The Korea Times*, 19 May 2009. [Electronic]. Available: http://koreatimes.co.kr/www/news/nation/2009/05/113_45190.html [24 November 2014].

Lee, H. (2014). Foreign visitors to Seoul exceed 10 million in 2013. *The Korea Herald*, 23 January 2014. [Electronic]. Available: http://www.koreaherald.com/view.php?ud=20140123000963 [24 November 2014].

Lee, J.Y. and Anderson, C.D. (2013). The restored Cheonggyecheon and the quality of life in Seoul. *Journal of Urban Technology*, 20: 3–22.

Limb, J. (2014). President attends groundbreaking for LG science park. *Korea Culture and Information Service*, 28 October 2014. [Online]. Available: http://www.korea.net/NewsFocus/Policies/view?articleId=122491 [9 December 2014].

McKeag, A. (2014). Freeway down! Seoul removing 16th freeway. *Congress for the New Urbanism*, 4 July 2014. [Online]. Available: http://www.cnu.org/cnu-salons/2014/04/freeway-down-seoul-removing-16th-freeway [23 November 2014].

Mesmer, P. (2014). Seoul demolishes its urban expressways as city planners opt for greener schemes. *The Guardian*, 13 March 2014. [Electronic]. Available: http://www.theguardian.com/world/2014/mar/13/seoul-south-korea-expressway-demolished [11 November 2014].

Park, W.S. (2013). Forging ahead with cross-sector innovations. *Stanford Social Innovation Review*. Summer. [Electronic]. Available: http://www.ssireview.org/articles/entry/forging_ahead_with_cross_sector_innovations [10 April 2015].

Pressian (2004). Housing demand 50% of supply. *Pressian*, 5 October 2004. [Electronic]. Available: http://www.pressian.com/news/article.html?no=28090 [30 October 2014]. (In Korean.)

Seoul Institute (2014). The role and function of urban think tank. *The Seoul Institute*. [Online]. Available: https://seoulsolution.kr/sites/default/files/notice/Seoul%20Institute_0.pdf [9 December 2014].

Seoul Metropolitan Government (SMG) (2012). *Knowledge Industrial Green City Leading the Future of Northeast Asia: Magok District*. Seoul.

Seoul Metropolitan Government (SMG) (2013). Seoul Metropolitan Government presents a blue print for the development of the MICE industry. *Seoul Metropolitan Government*, 13 November 2013. [Online]. Available: http://english.seoul.go.kr/seoul-metropolitan-government-presents-a-blue-print-for-development-of-the-mice-industry/ [9 December 2014].

Seoul Metropolitan Government (SMG) (2014). Smart Seoul 2015. *Seoul Metropolitan Government*. [Online]. Available: http://english.seoul.go.kr/wp-content/uploads/2014/02/SMART_SEOUL_2015_41.pdf [15 November 2014].

Seoul Metropolitan Government (SMG) (2015). *Seoul Station 7017 Project*. [Online]. Available: www.ss7017.org/en/index.jsp [21 March 2015].

Shwayri, S.T. (2013). A model Korean ubiquitous eco-city? The politics of making Songdo. *Journal of Urban Technology*, 20: 39–55.

Shwayri, S.T. (Middle Initial) (forthcoming). Walking back to happiness: Seoul towards a pedestrian-friendly core. *International Journal of Urban and Regional Research*.

Spilhaus, A. (1967). The experimental city. *Daedalus*, 96: 1129–1141.

The Hankyoreh (2007). Mayor aims to revive Seoul's fame as waterfront city. *The Hankyoreh*, 4 July 2007. [Electronic]. Available: http://english.hani.co.kr/arti/english_edition/e_national/220351.html [30 October 2014].

The Korea Times (2008). Magok district, an eco energy town of the future. *The Korea Times*, 24 June 2008. [Electronic]. Available: http://koreatimes.co.kr/www/news/issues/2008/06/139_26460.html [9 December 2014].

World History Project (1995). Sampoong department store collapse. *World History Project*, 29 June 1995. [Online]. Available: http://worldhistoryproject.org/1995/6/29/sampoong-department-store-collapse [23 January 2015].

Yoon, S. (2014). The Seoul Metropolitan Government announces a new urban architecture road map for the next four years. *E-Space*. 31 October. [Electronic]. Available: http://www.vmspace.com/eng/sub_news_view.asp?idx=6173 [25 November 2014].

14 Frankenstein cities

(De)composed urbanism and experimental eco-cities

Federico Cugurullo

Introduction: experimental eco-cities

Across the world, geographical studies show how the quest for the sustainable city has been interpreted and approached differently according to different contextual backgrounds (Whitehead 2003, 2007). However, among a number of heterogeneous planning strategies, it is possible to identify the contours of common urban trends. Over the last few decades, the construction of new cities has emerged as one of the leading strategies to pave the road towards a condition of urban sustainability. In less than ten years, over 30 projects for new master-planned sustainable cities have been developed in both the Global North and South, fuelling a global trend that is now shaping how sustainable urbanism is understood and practiced (Cugurullo 2013a). *Eco-cities* is how these new settlements have been broadly promoted by developers and stakeholders and categorized by urban scholars.

As shown in a number of studies, the eco-city phenomenon is a complex and diverse urban trend (Joss *et al.* 2011). First, the eco-city label often hides interests and objectives that are far from the tenets of sustainable development. Climate change mitigation, urban ecology and socio-environmental justice are some of the key themes that are regularly embraced by eco-city developers but rarely integrated once eco-city projects enter the implementation phase (Cugurullo 2013a, 2013b). As a result, the material incarnation of the original eco-city ideals is, in many cases, an 'ecological enclave': a place where scarce environmental benefits are unevenly distributed and accessible only to dominant groups (Hodson and Marvin 2010: 311; see also Caprotti 2014; Rapoport 2014). Second, eco-city initiatives are marked by contextual variations that are largely determined by the history, geography, economy and politics of the region where their development takes place. Consequently, aspects of eco-city developments, such as the architecture and design of the new settlements, vary according to specific geomorphological and climatic contexts, while the way eco-city developers understand and practice sustainability tends to be dictated by regional and/or national political economies (Chang and Sheppard 2013; de Jong *et al.* 2013; Cugurullo 2015).

While the eco-city, as a global urban phenomenon and planning model, is aptly understood in urban studies as being variegated and heterogeneous, single eco-city initiatives, due to their master-planned nature, are commonly perceived as organic projects shaped by a homogenous, top-down vision of the sustainable city.

In the literature and in the media, projects for new eco-cities such as Masdar City, Songdo and Tianjin, are portrayed as systematic *experiments* that seek to reach a condition of urban sustainability by following omni-comprehensive, authoritarian master plans (see, for instance, Vidal 2011). Like contemporary versions of Modernist master-planned cities such as Le Corbusier's Chandigarh and Niemeyer's Brazilia, experimental eco-cities are gradually imposing themselves in the collective imaginary as the cities of the future: ideal urban developments that will cast away the spectre of climate change and the fears of an environmental apocalypse by means of universal and unilateral visions of sustainability.

This chapter takes a different position and argues that eco-city projects are not organic initiatives, but rather heterogeneous experiments made up of a plethora of different sub-experiments that ultimately lead to the production of different and diverse areas and elements of the city. Furthermore, the chapter claims that the non-organic nature of experimental eco-cities is one of the main reasons why eco-city projects often fail from a sustainability perspective and are unable to meet their ambitious ecological, social and economic targets. This claim is unpacked empirically in the following three sections through the exploration of a single case study, Masdar City in the United Arab Emirates, whose experimental nature is deconstructed and critiqued to (a) reveal the various architectural, technological and economic experiments that compose the Masdarian enterprise and (b) show how such variegated aggregation of experiments undermines the sustainability of the new city. Theoretically, the chapter contributes to contemporary debates on the nature and dynamics of urban experiments by developing the concept of *(de)composed urbanism*, whose characteristics are illustrated using Mary Shelley's Frankenstein as a metaphor for experiments generated by the forced union of different, decaying parts.

The Masdarian experiment

In 2006, the emirate of Abu Dhabi began the development of what would become one of the most influential and controversial urban experimental sites of the twenty-first century: Masdar City. A US$20 billion public investment, Masdar City was designed by the international architectural studio Foster and Partners (F+P) as a six-square-kilometre settlement meant to reach a condition of sustainability by means of a mix of high-tech clean devices and traditional architecture. Presented by developers and stakeholders as the first zero-carbon city in history, Masdar City quickly polarized the opinions of urbanists and environmentalists. Within a few years, influential environmental organizations such as WWF (World Wide Fund for Nature) and IRENA (International Renewable Energy Agency) became keen supporters of the Emirati eco-city project while several urban scholars critiqued the discrepancies between the developers' promises and what was actually being built, especially with regards to the social and environmental aspects of the development, that appeared to be overshadowed by the economic interests of Abu Dhabi (see, for instance, Ouroussoff 2010; Cugurullo 2013a,).

Over the years, the concerns over the sustainability of Masdar City have been repeatedly confirmed by the work of geographers, planners and political scientists who have studied the materiality of the project and have unearthed an undemocratic urban experiment where lofty environmental claims hide bigger economic objectives (Crot 2013; Cugurullo 2013b). Recent empirical studies have demonstrated that the Masdarian experiment is economic in nature. The new city is used as an urban laboratory where clean-tech companies from all over the world develop, test and commercialize new products (Cugurullo 2013a). In the Emirati eco-city project, the production of space is subordinated to the production of capital and socio-environmental concerns are not taken into account unless they can be capitalized. Overall, in eco-city studies, there is widespread consensus that Masdar City has failed to establish a balance between economic, environmental and social interests, and instead has become a mere economic tool in the hands of the government of Abu Dhabi.

This chapter explores the Masdar City project by looking specifically at the planning dynamics through which the Masdarian experiment has been implemented. It argues that the experimental nature of Masdar City reveals why and how the city has failed to keep its environmental and social promises, thereby undermining its sustainability potential. The chapter is based on in-depth empirical research undertaken in Abu Dhabi from September 2010 to May 2011. Nineteen semi-structured and fifteen unstructured interviews were conducted with representatives from the developers of Masdar City, the Masdar Initiative, the architectural studio behind the master plan of the new city, F+P, and key business partners, Siemens, Schneider and Mitsubishi, involved in the implementation of the project. In addition, key documents including master plans and environmental reports were examined to triangulate the information that emerged in the interviews and produce extra data on the planning process and its ecological impact. Abu Dhabi is controlled by an authoritarian regime averse to critical political and urban research, and, thus, interviewees have been anonymized and are only identified by their roles in the Masdar City project.

X actors = X experiments

The main reason why Masdar City is not and cannot be a homogeneous urban experiment is that the project involves the participation of a variety of public and private companies and institutions that have different and often colliding interests and understandings of sustainability. This section aims to deconstruct the composition of the Masdar City project to identify (a) who the key actors are, (b) what they want to get out of the project, (c) how their interests shape the way they understand sustainable cities, and (d) how their interests and ideas of urban sustainability are translated into different urban experiments.

The actor leading the development of Masdar City is the Masdar Initiative, a state-owned company established in 2006 to research and develop clean energy solutions and technologies. The Masdar Initiative has two main aims. First, it has to help Abu Dhabi develop a new economic sector to prevent the economic and political collapse of the emirate in the upcoming post-petroleum era. Second, as a company financed entirely by the local government, it has to ensure that the Masdar

City project is profitable and that the construction of the new city generates capital that Abu Dhabi can use to sustain its economy. Consequently, the way the Masdar Initiative understands sustainability is largely in economic terms and, for it, a sustainable city is and has to be a city that generates profit for the foreseeable future.

The Masdar City project is supported by several private companies. Being a company targeting the development of new clean technologies, the Masdar Initiative has established a number of partnerships to finance, co-develop and commercialize its creations. Big multinationals like Siemens, Schneider, General Electric and Mitsubishi are among the key partners of the developers of Masdar City, together with several emerging companies keen to establish themselves in the clean-tech market. Profit is the top priority of these private companies whose understanding of sustainability is in sync with that of the Masdar Initiative. For them, sustainable development cannot be detached from economic development as the sale of products and services sustains their life.

Behind the design, architecture and planning of Masdar City is a team of architects and planners from F+P, the architectural studio that in 2007 won an international competition set up by the government of Abu Dhabi to identify the best master plans for a new eco-city. The studio has two contrasting objectives. Intellectually, it aims to design and build a low carbon settlement by merging traditional Islamic architecture and cutting-edge clean technologies. Professionally, it has to follow the request of its client, the Masdar Initiative, and ensure that the architecture of the new city supports the businesses in the city.

The above interests and understandings of sustainability do not have the same status. The Masdar Initiative, the actor representing the interests of the sole funder of Masdar City (the government of Abu Dhabi) frames the project and defines the basic structure of the Masdarian experiment. The formula underpinning the experiment is linear and simple: Masdar City is meant to be an urban space where the Masdar Initiative and its business partners can research, test and develop new clean technologies that eventually will be commercialized to generate profit (Cugurullo 2013a). This formula allows the Masdar Initiative to fulfil its key aim. The production of new clean technology feeds into Abu Dhabi's plan to create a new post-oil economic sector while the profit coming from the commercialization of clean-tech products generates capital that can be translated into regional and overseas investments, thereby increasing the emirate's global portfolio of financial assets which, to date, includes stakes in Barclays, Virgin Galactic and Manchester City Football Club, for an estimated total of US$300–$875 billion (The Economist 2008; Sovereign Wealth Fund Institute 2013).

The formula of the Masdar Initiative is in line with the targets of its business partners. For companies like Siemens, Mitsubishi and Schneider, an urban space like Masdar City represents a living laboratory where their new products can be tested in a real-life environment. Masdar City allows them to collect data on clean-tech devices such as smart grids and concentrated solar power stations that are integrated directly in the urban fabric of the city, thereby becoming a permanent feature of the settlement. Moreover, researching, implementing and testing new technologies in a real-life environment means building a competitive advantage over companies that develop their products in traditional indoor laboratories. This explains why clean-tech companies from all over the world have been eager to secure a spot in Masdar City.

For F+P's architects and planners, the situation is different. Their original master plan consisted of an entirely car-free city characterized by a mix of elements of traditional Islamic architecture and urban design, such as wind towers positioned within the grid of narrow streets meant to channel the wind and reduce the perceived air temperature in the city (see Figure 14.1), and elements of high-tech architecture such as an automated, driverless transport system designed to connect every area of the settlement. However, there is a key difference between the urban formula of the Masdar Initiative and F+P's plan, inasmuch as the former is meant to be fluid and adaptable. The Masdar Initiative has to keep the Masdarian experiment flexible to accommodate the interests of its business partners. As explained above, the physical structure of the city changes every time a company develops

Figure 14.1 Masdar City's wind tower (source: author).

and installs a new technology. For example, if a company like Schneider decides to invest in roof-mounted solar panels this means that some of the roofs of the buildings in Masdar City will have to be designed to accommodate the installation of that specific technology. And these changes can be significant. The development of a smart grid, for instance, would imply the redevelopment of the entire surface and undercroft of Masdar City to allow the passage of cables and fibre channels, while the development of a concentrated solar power station would require the developers to build a whole new area.

As clean-tech companies are always looking to develop new products that differ from what is already on the market, it is impossible for the Masdar Initiative to foresee what will be developed and implemented in Masdar City and how the city will change accordingly. Therefore, in Abu Dhabi's new city, no precise master plan and vision of sustainability is possible. This has a drastic impact on the socio-environmental performance of the settlement, as explained in the next section.

X sustainabilities = 0 sustainability

The nature of the business that underpins the Masdar City project brings together a number of actors that use the city to experiment with new urban technologies. However, being aleatory processes, the outcomes of experiments can be negative and so can their socio-environmental impact. The development of Masdar City's Personal Rapid Transit (PRT) exemplifies the fragility of the experiments taking place in the new Emirati city. The PRT was one of the key elements of the original master plan developed by F+P in 2007. Engineered to be connected to an overarching smart grid, it aimed to provide free public transport across the settlement. Physically, it consisted of a series of cars powered by renewable energy: the driverless cars follow sensors integrated in the urban fabric of the city after passengers select their destination on a touch-screen inside the vehicle. The project had a double socio-environmental purpose. Socially, it sought to eliminate automobile dependency and encourage people to walk and use public services, to promote well-being and community cohesion. Environmentally, the objective was to decrease the carbon impact of the settlement, as the PRT system was designed to connect spaces via the shortest route and be supplied by a smart, zero-waste production and circulation of clean energy.

Between 2008 and 2010, the PRT was implemented in approximately 10 per cent of Masdar City, linking the entrance of the city to its centre. Environmentally and socially, the experiment was successful, but not economically. In 2010, when the global financial crisis hit Abu Dhabi, the local government decided to reduce the investments on the new city and the Masdar Initiative had to rescale the project and abandon the implementation of its most expensive features, including the PRT. Since then, while it is still possible to use the PRT to reach the centre of the settlement, Masdar City has opened its doors to electric cars. From a planning perspective this meant the death of F+P's experiment for the first fully pedestrian city in contemporary history, while for the Masdar Initiative and Mitsubishi (the company leading the development of the Masdarian electric cars) it represented the beginning of a new one.

The story of the PRT is not unique. Since it joined the Masdar City project, Siemens has launched over 50 projects, including smart grids, smart sensor networks and photovoltaic power systems, but it is now developing only a fraction of the original portfolio. According to F+P's architects, the constant turnover of business partners and clean-tech projects has been a major problem in relation to the original socio-environmental ambitions of Masdar City. There is little to no coordination among the different techno-urban experiments that take place in the new city, and even when the architects disagree with the approach to urban sustainability of a company, they cannot reject it as, in Masdar City, economic interests come first.

'The process can be frustrating' confessed an architect from F+P in an interview. According to him, in 2007, when the studio was conceptualizing the project, the biggest intellectual challenge was to come up with a cohesive master plan for a zero-carbon and zero-waste city meant to be built in one the most environmentally challenging places in the world: the Emirati desert where the temperature can reach 49°C in the summer. The challenge was addressed by looking at cases of existing cities developed in similar climatic conditions. Shibam (Yemen) and Aleppo (Syria) with their compact design and structure provided the source of inspiration that led F+P towards the idea of a high-tech, hyper-compact settlement where, largely thanks to traditional architecture, the perceived air temperature is ten degrees lower than in the rest of Abu Dhabi. Today, the challenge is of a different nature. As the architect explained, Masdar City is a 'patchwork' composed of different pieces of urban fabric produced by different clean-tech projects. The challenge is to 'plug-in' all these different elements while preserving the sustainability of the city.

It is no mystery that, at the time of this writing, F+P is failing to save the sustainability potential of Masdar City. In 2010, in light of the poor environmental performance of the settlement, the Masdar Initiative changed the sustainability targets of the new city, replacing its promotion as a zero-carbon settlement in favour of a more feasible 'carbon-neutrality' target. In essence, the development of Masdar City produces too much technology and urban space for the city's clean sources of energy to support. The technological ambitions of the Masdar Initiative and its business partners have negated the ecological ambitions of F+P. To sustain their techno-urban development, the developers have to rely on off-site sources of energy and, more specifically, on oil and gases to produce the electricity to maintain the Masdarian experiment. Indicative of the environmental failures of Masdar City is also the fact that since 2007, the Masdar Initiative has published only one environmental impact assessment (EIA). Published in 2009 by Hyder Consulting, the document and the analysis that underpins it were produced when the implementation of the project was in its very early stages. Consequently, the assessment does not evaluate the materiality of the settlement and its ecologic impact. Instead, it focuses on the urban future of Masdar City, speculating over what the settlement might become in the next 20 years. Since then, no EIA has been released and developers and stakeholders have carefully avoided disclosing any data on the environmental impact of the settlement, keeping the results of their experiments confidential.

Conclusion: (de)composed urbanism

> His limbs were in proportion, and I had selected his features as beautiful. Beautiful! Great God! His yellow skin scarcely covered the work of muscles and arteries beneath; his hair was of a lustrous black, and flowing; his teeth of a pearly whiteness; but these luxuriances only formed a more horrid contrast with his watery eyes, that seemed almost of the same colour as the dun-white sockets in which they were set, his shrivelled complexion and straight black lips.
>
> (Shelley 2013: 45)

These are the words through which Doctor Victor Frankenstein describes the results of his experiment. The creature is composed of several different pieces that the doctor had assembled to create the perfect being. Every piece had been carefully selected, as Frankenstein's plan was to create a creature that was not only immortal, but also beautiful. The results, however, are dreadful and the appearance of the experimental creature is that of a monster. What is notable in the words of the doctor is that there is nothing wrong with the single pieces. Muscles, hair and teeth are close to perfection: it is the *contrast* that makes Frankenstein's creature a monster. It is the heterogeneity of the elements that the doctor forces together into a single artefact that eventually compromises the experiment.

In a similar vein, Masdar City is the product of a process of assemblage that connects different pieces of urban fabric into a single settlement. Behind every single element are meticulous calculations and studies, and, individually, most of the components of the new city work as their creators had hypothesized. Smart grids circulate data across the settlement, wind towers and narrow streets channel the wind to reduce the perceived temperature, concentrated solar power stations produce clean energy, and electric cars and automated vehicles facilitate transport and mobility. However, as this chapter has shown, there is no homogeneous vision and strategy of sustainable urban development linking these elements together and, ultimately, the contrast created by the juxtaposition of the different parts of the city affects the overall sustainability of the settlement.

Masdar City is an example of what this chapter defines as *(de)composed urbanism*. The new city is the result of different clean-tech and architectural experiments whose products are assembled to create what is misleadingly promoted as an eco-city. As such, Masdar City is composed of a plethora of parts that do not organically feed into an overarching sustainability strategy, thereby compromising the original vision of the architects and planners behind the master plan. Moreover, some of the elements are generated by failed experiments, such as the PRT, that developers and stakeholders have abandoned. Nonetheless, although 'dead' from a planning perspective, these parts of the settlement are kept alive via off-site energy sources (oil and gases) to prevent their decomposition.

Putting the case of Masdar City into historical perspective shows that there is nothing new with the fragmented planning strategies and practices observed in Abu Dhabi's urban experiment. Medieval cities, for example, as Benevolo (1993) points out, were shaped by an irregular and often chaotic layout symptomatic of the different interests of the multitude of actors behind the politics of the city. In the medieval city, the fragmented nature of urban spaces reflected the fragmented nature of political and economic power that was unevenly distributed amongst noblemen, religious orders and merchants (Mumford 1961; Benevolo 1993). Since the Middle Ages, the stakeholders have changed as bishops, princes and guilds have been replaced by actors such as states, corporations and non-governmental organizations, but the shape of a city continues to reflect the shape of the different powers that rule it politically and economically. What differs is that in the twenty-first century, science has produced an understanding of the environmental impact of urbanization and raised awareness about the need to carefully plan and build cities sensitive to local and regional ecologies and climates. By examining the development of Masdar City, what this chapter has shown is that fragmented planning strategies do not promote urban sustainability. The absence of an organic set of actions that systematically and analytically study, assess, direct and adjust the urbanization of an environment from a socio-ecological angle is the missing element within the Masdarian machine: an element that is crucial not only to eco-city projects but to all urban experiments including smart cities and regeneration initiatives that target the formation of sustainable built environments.

However, it is important to remember that the practice of *(de)composed urbanism* is not the only reason why new cities like Masdar City manifest poor social and environmental performances. The Emirati project for a new eco-city comes from a specific political-economic rationale that seeks to sustain economics and politics rather than the environment. While F+P had envisioned an eco-city as a low carbon, climate-sensitive settlement, the Masdar Initiative, as a state-owned company, had envisioned a test bed to develop and commercialize clean-tech products with its business partners. In this sense, the Masdarian experiment fails, environmentally and socially, because, in the mind of the developers, it was never meant to succeed. The negative environmental and social externalities generated by the development of Masdar City are the product of an economic experiment: the only experiment that the government of Abu Dhabi has an interest in.

Like Frankenstein's monster, Masdar City is an artefact that hides itself, seeking to conceal any information on its environmental performance. The exact fate of this urban creature is hard to predict, but it is possible to foresee some of the socio-environmental repercussions of the experiment. A city that is artificially kept alive by off-site, finite energy sources and governmental investment sustained by the petroleum industry cannot be immortal and is meant to collapse due to its scarce environmental and social foundations. In Mary Shelley's novel, the creature is abandoned by its creator and dies alone. Today, Masdar City is still under construction and, although unlikely, a positive finale is still possible.

References

Benevolo, L. (1993). *The European City*. Oxford: Wiley-Blackwell.

Caprotti, F. (2014). Eco-urbanism and the eco-city, or, denying the right to the city? *Antipode*, 46: 1285–1303.

Chang, I.C.C. and Sheppard, E. (2013). China's eco-cities as variegated urban sustainability: Dongtan Eco-City and Chongming Eco-Island. *Journal of Urban Technology*, 20: 57–75.

Crot, L. (2013). Planning for sustainability in non-democratic polities: the case of Masdar City. *Urban Studies*, 50: 2809–2825.

Cugurullo, F. (2013a). How to build a sandcastle: an analysis of the genesis and development of Masdar City. *Journal of Urban Technology*, 20: 23–37.

Cugurullo, F. (2013b). The business of utopia: Estidama and the road to the sustainable city. *Utopian Studies*, 24: 66–88.

Cugurullo, F. (2015). Urban eco-modernisation and the policy context of eco-city projects: where Masdar City fails and why. *Urban Studies*, earlyview online.

de Jong, M., Wang, D. and Yu C. (2013). Exploring the relevance of the eco-city concept in China: the case of Shenzhen Sino-Dutch Low Carbon City. *Journal of Urban Technology*, 20: 95–113.

The Economist (2008). Asset-backed insecurity. *The Economist*, 17 January 2008. [Electronic]. Available: http://www.economist.com/node/10533428 [25 September 2015].

Hodson, M. and Marvin, S. (2010). Urbanism in the anthropocene: ecological urbanism or premium ecological enclaves? *City*, 14: 298–313.

Joss, S., Tomoseiu, D. and Cowley, R. (2011). *Eco-Cities – A Global Survey 2011*. [Online]. Available: http://www.westminster.ac.uk/?a=119909 [30 January 2015].

Mumford, L. (1961). *The City in History: Its Origins, Its Transformations, and Its Prospects*. New York: Harcourt Brace.

Ouroussoff, N. (2010). In Arabian desert, a sustainable city rises. *The New York Times*, 25 September 2010. [Electronic]. Available: http://www.nytimes.com/2010/09/26/arts/design/26masdar.html?pagewanted=all&_r=0 [30 January 2015].

Rapoport, E. (2014). Utopian visions and real estate dreams: the eco-city past, present and future. *Geography Compass*, 8: 137–149.

Shelley, M. (2013). *Frankenstein: Or, the Modern Prometheus*. London: Penguin.

Sovereign Wealth Fund Institute (2013). *Sovereign Wealth Fund Rankings* [Online]. Available: http://www.swfinstitute.org/fund-rankings/ [30 January 2015].

Vidal, J. (2011). Masdar City – a glimpse of the future in the desert. *The Guardian*, 26 April 2011. [Electronic]. Available: http://www.theguardian.com/environment/2011/apr/26/masdar-city-desert-future [30 January 2015].

Whitehead, M. (2003). (Re)analysing the sustainable city: nature, urbanisation and the regulation of socio-environmental relations in the UK. *Urban Studies*, 40: 1183–1206.

Whitehead, M. (2007). *Spaces of Sustainability: Geographical Perspectives on the Sustainable Society*. London: Routledge.

15 Experimental afterlives

Making and unmaking developmental laboratories in Ghana

Thomas Yarrow

Introduction

This chapter traces the after-effects of an experiment in urban planning. In the wake of the construction of the Akosombo Dam, following Ghana's Independence in 1957, 80,000 people were resettled under an ambitious scheme of planned development that aimed to turn rural peasants into modern citizens through the creation of model townships. As in other mid-century resettlement schemes, the language of 'experiment' was itself integral to the project. By the 1960s, the paradigm of 'science for development' had been largely superseded by an understanding of development as an experimental science oriented towards the 'improvement' of living conditions through the application of expert forms of knowledge (Bonneuil 2000). As an instance of a broader experimental logic of development, the Volta Resettlement Scheme aimed to redesign rural life and became an important site for the production and enactment of various forms of expertise. In this project (Shapiro 2003; Miescher 2012) and a variety of other post-colonial contexts (e.g. Holston 1989; Mbembe 2004; Roy 2007), the desire to be modern became axiomatically linked to the desire to be urban. Urbanisation was equated with development as a means of rendering rural populations more amenable to state-led industrialisation and as the embodiment of modernist aspirations of efficiency and rationality in their own right. Framed by broader ideological currents, the post-independence Nkrumah government presented the project as a literalisation of aspirations for the rapid achievement of a socialist pan-African modernity that would realise the benefits of western development while obviating its associated problems through careful planning (Shapiro 2003). From the outset, resettlement townships were imagined as exemplary spaces and were powerfully invested with a sense of future possibility. As experimental sites, they were important less for what they were than for what they, as vanguards for the nation, would one day become.

Occupants of these townships today take the built environment as evidence of the failure of this experiment but continue to maintain their faith in the visions that inspired it. Images of urban modernity drawn from 1960s planning discourses intersect with a broader set of discourses about the urban and the modern, orienting the ways in which resettlers imagine and act upon the built environments they inhabit. Urban modernism is thus constituted in a powerful yet ineffable sense of felt absence and in an ontological orientation that constitutes these spaces through a series of negative contrasts in terms of what they are not. The material remains of these experiments in resettlement evoke for their inhabitants people and processes that failed to materialise as hoped. The collapsing and decaying fabric of their houses recalls collapse of the vision that originally animated their construction and elicits competing accounts of the history that led to this failure. Abandoned and ruined houses literalise an absence of the development that was promised. Lacking the money for cement, houses have been extended in local materials, leading to a proliferation of unplanned structures whose explicitly 'temporary' nature references a sense of arrested development and of a present in uncertain relation to the experimental futures of the past.

As an experiment in urban planning, the project was oriented towards the achievement of urban modernism and therefore anticipated a future that has failed to materialise in the terms imagined either by the officials and experts involved or by resettlers themselves. Yet the experiment left material and ideological legacies that remain imbricated in the lives of those who inhabit these townships today. This chapter is based on ethnographic work in two of the resettlement townships, including archival research and interviews with planners and other experts. I start by tracing the experimental logics that were central to the project's inception, before examining how these persist both as a set of now decaying material infrastructures, and as a series of aspirations and ideologies. Through this, I highlight the specific disjunctions that emerge between utopian framings of experimental futures and the infrastructures of the projects involved.

The sociologist Borup has noted how new technologies foster an historical amnesia: 'hype is about the future and the new – rarely about the past – so the disjunctive aspects of technological change are often emphasized and continuities with the past are erased from promissory memory' (Borup *et al.* 2006: 208). While as much was true of the Volta Resettlement Project at the time of its inception, Africa is now undergoing a range of experiments in urban and low carbon living (Silver and Marvin 2015) that produce a similar amnesia with respect to urban experiments of the past. In this context, the lens of the Volta Resettlement is useful as a reminder that present experiments are configured in relation to an infrastructural and ideological legacy of projects that persist, more or less palpably, even in manifest failure. New experiments are built not only on the crumbling infrastructures of mid-century large-scale development schemes but also on the aspirations and ideologies these earlier experiments set in train. Thus, my focus on the material and ideological afterlife of experimental infrastructures rejoins recent work on post-colonial ruination to highlight how pre- and post-colonial formations remain visibly and viscerally present in the materiality of what (literally) remains (Stoler 2008). In conclusion, I suggest that the particularities of the Volta Resettlement in turn provide a lens that makes apparent some of the analytic limitations of recent work on ruination.

Resettlement as experiment

The Volta Resettlement Project was undertaken in the wake of the construction of the Akosombo Dam during the early to mid-1960s. Although plans for the dam can be traced back well into the colonial period, the project was finally initiated following Ghana's Independence in 1957. As a flagship policy of Kwame Nkrumah, Ghana's first president, the physical construction of the dam became central to the construction of the newly independent nation of Ghana in ways that were indissolubly material and symbolic (cf. Mitchell 2002). The project entailed a package of linked changes that were imagined to set in train progress to a different and better future via a break with the traditional pasts of displaced peasants. Key elements of this high-modern approach were the promotion of social and economic development through the creation of planned urban spaces, technological modernisation of agriculture, and technologically driven industrialisation, notably linked to the power produced by the dam. This ideology drew explicit inspiration from other socialist contexts, specifically in an understanding of planned urban infrastructure as generative of modes of transformative modernity (Holston 1989, Alexander and Buchli 2007)

In various government discourses and in wider media coverage, resettlement was seen as a 'sacrifice' made by these inundated communities on behalf of the nation as a whole but also as a form of development and improvement. Following completion of the dam in 1966, an editorial in the state owned *Ghanaian Times* (1966: 5) praises the selflessness of inundated communities suggesting, 'Not least in the pride of place of honour and praise are those Ghanaians whose love for the motherland and the prosperity of mother Africa, sacrificed their lot to bring the project to fruition. History will not forget them.' If resettlement was seen to entail 'loss' in terms of particular traditions and the disruption of 'organic' if 'backwards' communities, politicians, planners and government officials suggested that these would be outweighed by the corresponding improvements associated with the promised urban modernism. In a widely reported quote, frequently retold by inhabitants of the resettlement communities today, Nkrumah made a personal pledge that none of the resettled communities would be worse off as a consequence of the move. As the subsequent discourses of planners, architects, journalists and politicians testify, the intention was for resettlement to bring about rapid and sweeping improvement, specifically through the construction of 'urban' and 'modern' townships.

The project was explicitly experimental, in a number of related respects (Shapiro 2003). The scientific approaches to development that pervaded in the 1960s both assumed and produced a vision of society as an object with its own laws and which could be worked on through technical procedures. These, as Paul Rabinow observes in the context of Morocco (1989), were becoming the authoritative arbiters of what counted as 'real'. Development as an experimental science thus entailed a belief in the possibility of a different, better future – 'improvement' and 'development' – through the application of various forms of scientific knowledge to social and economic problems. A paper by D.A.P Butcher (1970: 88), who oversaw the social survey, makes evident the kind of assumptions that prevailed: 'Resettlement is largely a problem in human engineering and, as such, the "engineers" – in this case the Social Welfare Officers – have to keep track of their materials, which, being people, will not keep still like cement and iron.' If society was a system governed by universal principles, its management required the kind of expertise that social scientists were able to provide.

The importance of these townships as experimental sites necessitated the physical and conceptual bounding of resettlement townships from the surrounding villages. Their national significance arose not from their typicality but rather in their distinctiveness from the social and economic characteristics of other parts of rural Ghana and from the villages from which the resettlers had been displaced (cf. Roy 2007). The scheme involved a range of experts including planners, architects, engineers and agronomists, including Ghanaian elites – many educated abroad as part of the colonial governments' plans for Independence – and expatriates from Europe and America. As 'living experiments' (Shapiro 2003), the aim of the project was not only to transform the lives of the 80,000 resettlers but to derive planning and design principles that would form templates for a broader process of national transformation.

Bonneuil (2000) has argued persuasively that during this period, the logic of scientific experimentation was intricately implicated in the functioning and expansion of the state in various African contexts. Experimental discourses made authoritarian and productivist interests of colonial and post-colonial governments appear as a form of intellectual progress through the pursuit of knowledge. Where money was wasted or anticipated benefits failed to accrue, these could be justified as instances of experimental failure that did not undermine – and even ultimately bolstered – broader ideals of scientific truth and progress.

Resettlement in Ghana, as elsewhere, helped to bring about a shift in the relation between government and rural populations. Previously beyond the reach of the state, resettlement entailed forms of geometricisation, standardisation and discipline, shaping agrarian societies to make them more amenable to intervention and control. Small-scale peasant farmers were aggregated in larger, more 'urban' units that could be more easily administered. Bonneuil (2000: 269) notes in relation to the continent as a whole that 'village layout and housing as well as social life were also designed from above, so as to turn villages into functional units of supervision and experimentation'. Scientific experimentation and increasing state control went hand in hand insofar as both depended on and created similar forms of legibility and visibility:

> One can view the settlement schemes as crucial sites for aligning rural societies with the conditions and practices of the [research] station … They were hence 'experimental systems' … in so far as they constituted an arrangement of objects and people designed to produce experimental data.
>
> (Bonneuil 2000: 272–273)

The infrastructures that accompanied the Volta Resettlement were experimental, ordering the conditions of social and economic life in ways that made control, and hence measurement, possible. They were also experimental in the additional linked sense that they deliberately manipulated events to an anticipated but unknown outcome. In the context of the Volta Resettlement Project, urbanisation and experimentation were thus linked in two key respects. Planned townships provided the material infrastructures that made experimentation possible through order and legibility. At the same time, visions of urban life configured ideas about an anticipated future whose realisation was experimental and hence uncertain.

By bringing together previously distinct villages in planned townships, planners aimed for a more efficient use of resources as services were centralised in accordance with the tenets of Central Place Theory, an internationally fashionable modernising planning discourse influenced by the work of Walter Christaller. Under the Volta Basin Area Development Plan, hundreds of small subsistence villages, were aggregated into 54 townships. As well as enabling the centralisation of services such as schools, clinics, water pumps and public toilets, the agglomeration of previously distinct communities was intended to provide a compact labour force that would facilitate the mechanisation of agricultural production and provide a further impetus to economic growth. In contrast to the inundated communities, the regional plan was intended, in the words of the Chief Planning Officer 'for creating a more rational and economic pattern of settlements', in which 'the selection of suitable sites, the pattern of settlements and their sizes will be influenced by such technical considerations as health, sanitation and water supply'.[1]

If the logic of experimentation – and more broadly of science – participated more or less wittingly in the development of new forms of governmentality, these experimental spaces also created 'laboratories' through which new knowledge, new paradigms of thinking and even new disciplines were founded (Bonneuil 2000). For researchers from a range of disciplines including sociology, anthropology, engineering, architecture, planning and agronomy, resettlements made people and the environments in which they lived increasingly accessible and legible, providing a context in which various forms of intervention could be undertaken and monitored. The townships were thus differently constituted as experimental sites in relation to a range of epistemic interests. Resources from the Volta Resettlement Project, and the experimental possibilities this made available, helped establish a number of new university departments, through work undertaken at Kumasi and Legon. The Buildings Research Group at the Kwame Nkrumah University of Science and Technology played a leading role in the development of building technologies that sought to combine modern principles with locally available materials and a sensitivity to existing social circumstances. A social survey, undertaken to aid the successful integration of different groups, contributed to the development of new methodologies (pioneering an early IBM punch-card system for analysis of data on an unprecedented scale). Likewise, anthropological work in these communities contributed important new insights and helped develop new frameworks for the analysis of rural communities undergoing processes of 'modernisation'. If these kinds of knowledge participated in the emergence of new forms of governmentality, the academics who undertook this work were often explicitly critical of the government officials and the top-down methods espoused. Robert Chambers was involved in social research connected to the Volta Resettlement Project and edited the book that brought together key strands of the research supported by the project. His unease with the lack of community consultation is evident in his introduction to the book (Chambers 1970) and it is certainly possible that his later pioneering work on 'participation' (Chambers 1983) was at least in part built upon knowledge gained through this non-participatory process.

From the outset, the aims and ideologies that informed these experiments were far from monolithic, even amongst the various experts responsible for its design and execution (Miescher 2012). Tensions emerged regarding the extent to which such spaces should be configured to conserve or to change the existing lives of resettlers, the degree to which 'tradition' should be accommodated or overturned, and the emphasis to be given to pragmatic as opposed to idealistic visions of what could be achieved. Initial plans for resettlement were relatively modest, entailing the relocation of villagers through 'aided self-help'. However, the scale and scope of the plan increased as the project evolved, reflecting the increasingly ambitious developmental rhetoric of the newly independent government. As the lake waters started to rise, planners and politicians became increasingly concerned with the difficulties of resettling such a large number of people through self-help alone, particularly given the limited time available. Correspondence between planners, architects and civil servants testifies to an increasing desire to 'modernise' these communities through a process of planned urbanisation. In a memorandum produced by the senior assistant secretary to the Volta River Secretariat (Mensah 1961), he notes 'it is no good pushing Ghana 100 years back by giving the people the same old inferior buildings'. This change of emphasis is reflected in the disappearance of the word 'village' from the vocabulary of those implementing the resettlement plans, and its replacement by the word 'town' (Chambers 1970).

If the increased size of communities was desirable for the modernisation of economic activity and the rationalisation of services, the creation of a more 'urban' feel to the townships was also regarded as an end in itself. Commenting on the designs of the 'core houses' of which the resettlements were composed, Miles Danby (1970: 170), the British architect responsible for the design of one of these, positively remarks that 'this type gives a higher density … and had been used by planners to give a more "urban" or "town" feel'. Over and above these 'rational' considerations, the merging of ethnically and linguistically diverse populations was also seen as a positive move, facilitating a movement away from 'traditional' affiliations of kinship, chieftaincy and ethnicity towards more overtly 'modern' forms of relationship based on shared nationality and citizenship. In this way, resettlement townships were conceptualised as functionally integrated units in which social, ethnic and economic differences were subsumed to a regional and national logic of development.

Whilst government discourses foregrounded human agency and the capacity to shape nature to human ends, the realisation of these plans depended on complex and precarious alliances between a range of human and non-human agents (Mitchell 2002). Planning files testify to the unruliness of the people and things they sought to change: buildings cracked, materials were routinely stolen, cement caked, local workers failed to turn up and produced shoddy work, and resettlers retained habits and 'traditions' understood to be contrary to the logic of the modernist plan. The Volta Resettlement Project was from the outset precarious and contradictory, holding out a multiplicity of promises even in its apparent singularity.

I have pointed to the ways in which planners in post-colonial Ghana naturalised a relationship between the urban and the modern as intrinsically linked social forms. Such discourses constituted a set of aspirations that were concretely embedded in a range of infrastructures, including houses, planned townships, new roads and agricultural systems. These literalised a new relationship between citizens and state and carried forwards a set of hopes and expectations about the possibility of a different and better future. While this future has manifestly failed to materialise in the terms imagined, it leaves important material and ideological legacies. The next section explores the afterlife of these experimental spaces. I use the term 'afterlife' borrowing from Benjamin's conceptualisation of ruination in his work on the arcades projects (Dawdy 2010; Gordillo 2014). For him, as for other critical theorists, modern ruins act as reminders of the hubris of modern linear time, revealing its contradictions particularly powerfully as a consequence of their effaced functionality. The perspective provides a useful position to consider how experimental infrastructures continue to literally matter, even and indeed because they have 'failed'.

Experimental afterlives

Though the resettlement took place over 50 years ago, beyond the memory of the majority of residents of resettlement townships, ideas of urban modernity embodied in planning and public discourses of the 1960s remain central to the understandings and practices through which people today occupy these spaces. Resettlers frequently complain about the conditions within resettlement communities but rarely question the visions that informed these. High-modern developmentalist ideologies integral to the project's inception constitute a lens through which existing conditions are seen and found wanting. Assessments of today's circumstances relative to the standard of living prior to resettlement remain contested. However, resettlement communities are characterised by a pervasive sense of present conditions in a disjunctive and hence 'failed' relationship to the visions and promises that accompanied the project's inception. As an experiment in the development of a specific form of modern urbanism, the project entailed the promise of a future that persists in the memory of inhabitants of these townships and which animates a range of engagements with the now decaying infrastructure of what is widely seen as a failed experiment.

Abstractly, this absence is understood in terms of 'development', 'modernity', 'progress', 'civilisation' and 'urbanisation'. More concretely, such ideas are literalised in various aspects of the built environment. Buildings in poor repair are described as 'ramshackle' and 'un-civilised'. The 'temporary' nature of makeshift kitchens and bathrooms, fashioned from 'swish' (the local term for mud-constructed houses) and corrugated iron, is similarly highlighted as evidence of a gap between vision and reality (see Figure 15.1). More generally, the 'bushy', 'weedy' and 'chaotic' nature of the town are imagined as evidence of 'backwardness' and 'under-development', that highlights the failed promises that attended resettlement.

Such discourses evoke an understanding of ruination and decay that is partly elicited by the material remains of these infrastructures but which do not deterministically arise from these (Stoler 2008; Edensor 2012; Johnson 2013; Schwenkel 2013). Rather these relate to an ontology that enlists archetypes of the modern and the urban, and as the flip side of these visions, sees the environment to hand in terms of a relative deficit of those attributes. The physical remains of the resettlement project emerge in shifting disjunction with the ideological remains of visions that originally sustained it.

Walking around the resettlement township of Senchi with one of the town elders, he pointed out a core house that was being undercut by heavy erosion, causing it to sag heavily on one side: 'there is no development here, we are living like animals', he commented with frustration and despair. The house exemplified a wider predicament. In many of the resettlement townships, outward migration to larger urban conurbations and to 'home-towns' in other parts of the country have led to the abandonment of large numbers of the core houses that resettlers were allocated. Many of these structures now lie empty. Built in anticipation of future development by their

Figure 15.1 A 'core-house', typical of those provided to resettlers throughout the 54 resettlement townships. Note the front porch, intended for infill as an additional room. A separate kitchen has been added to the side. (Source: author.)

occupants, these structures now appear as empty shells that are seen to literalise an absence of the 'modernity' and 'development' they initially stood for. Materially deteriorating infrastructures, including collapsed walls, rusted tin roofs and rotting doors and windows symbolically index a wider sense of decay (cf. Stoler 2008). As literally empty structures, their material traces evoke the metaphorical absences of the promised modernity and the emptiness of the promises of planners.

Resettlers come to see their own lives through the lens of these modernist visions, which, in their manifest disjunction from contemporary realities, inspire both hope and despair. Modernist visions continue to circulate in nostalgic recollections of Nkrumah's plans, kept alive through various forms of oral history and memory that engender widespread consciousness of the project's promised futures. Born in Awura Hae, a small village resettled to New Senchi, Nana moved to Accra to work as a security guard before returning to be installed as a local chief. Now in his sixties, he lives in a small 'core house' adapted for the purposes of his role through the construction of a shrine and a small picket fence. Bemoaning the inadequacies of his 'palace' for the purposes of his office, he connected his impoverished status to the wider problems of the resettlement townships: 'If we were to be during Nkrumah's time, by this time Akosombo here would be very different … We don't get anything after Nkrumah's time.' Imagining an alternative trajectory in which Nkrumah remained in power, he described an alternative that in turn evoked the problems of present realities: 'If it was Nkrumah, you would find it difficult to enter my palace. You see, I will be growing tall and fine. But now, see, I am poor in everything.' Echoing these sentiments, an elderly man in the resettlement township of Npakadan linked the townships lack of development to the overthrow of Nkrumah:

> Life would have been really good – happier. Today we would have had an aerodrome, and Nkrumah also thought of making some rail lines from Kpong – many, many good things … The minute Nkrumah died all his plans and all that he wanted to do for the resettlement was stopped … So that is why we are suffering. Other than that, this town would be a very nice town.

Such sentiments articulate a broader sense in which present realities remain framed by the anticipated but unrealised promises of planners. The now crumbling remains of resettlement continue to be haunted by the alternative possibilities of the plans that initially prompted their construction.

Current social practice also acts to remake the modernist planned spaces that resettlement created. In various ways, the activities of resettlers exceed the possibilities that planners intended (Lefebvre 1991). Their day-to-day activities literally and conceptually extend these built forms in a range of ways that more or less subtly transform the modernist logics through which they were initially conceived. Planners sought to instate 'zones' in which 'business', 'commercial', 'domestic' and 'recreational' activities would be spatially segregated. In this way, the townships were imagined as a patchwork of areas in which social and economic 'function' coincided with particular spatial and built forms (Holston 1989; Lefebvre 1991). Such spatial distinctions are obviated through a range of practices.

The designation of resettlement houses as 'domestic' spaces is routinely transformed in the uses to which these houses are put. Core houses are extended and opened up onto streets so that they can function as shops, whilst others turn their houses into sites of commercial activity as makeshift 'chop bars', hair salons and bakeries. Spare rooms are used to store surplus produce such as corn and yams.

Architectural plans reveal the towns as a series of discrete 'plots' laid out along a regular grid system of interconnecting roads. The movements and activities of resettlers confound this segregation of space into 'public' and 'private' areas. Over the intervening four decades, many of these public rights of way have been eroded through illegal development or fallen into disrepair through lack of use. Whilst public rights of way have thus been effectively privatised, private space has also been made public. Commenting on the series of paths that cross-cut the township, a man living at the resettlement township of New Senchi joked to me, 'everyone's home is a path'. Architectural drawings depicting neatly trimmed boundary hedges thus depict a form of spatial segregation that in practice is rarely maintained.

If resettlers have thereby dissolved many of the distinctions that planners hoped to enact, they have also imposed socio-spatial distinctions that planners sought to erase. Planners, architects, resettlement officers and state bureaucrats saw resettlement entailing a movement from a 'traditional' past to a 'modern' future. Today, by contrast, resettlers understand the resettlement townships as a composite of temporally distinct elements. Houses, along with other aspects of the built environment, are described in terms of their relative 'modernity' or 'under-development' (Yarrow 2011). Spatial distinctions are thus presented in terms of temporal distinctions that in turn make apparent differences between the various groups of people of which the township is composed. In this sense, modernising narratives continue to inflect resettlers' understandings of these spaces, even as their use of these terms calls into question the homogeneous, empty time of the modern nation state.

In these and other ways, the daily practices of resettlers have acted to reassert social processes and cultural values that the plan intended to deny (cf. Holston 1989). What resulted was not an 'old' way of life, still less the hangover from tradition that planners and town managers have subsequently claimed. Neither, however, was it the imagined urbanism of planners and architects. Existing theories of modernist planning help to reveal how the production of the 'abstract space' (e.g. Lefebvre 1991) of planners and architects is implicated in the consolidation of new forms of state power and governance. By the same token, these illuminate how space is literally produced through a range of social practices that confound the logic of urban modernity that planners and architects intended to enact (de Certeau 1984). Yet in order to properly capture the dynamics at play in these resettlement townships, it is necessary to qualify these theories with respect to their formulation of the political (Nielsen 2011).

Conceived as an instrument of state power, it has become increasingly common to celebrate if not romanticise practices and relationships that appear to deny, evade or reformulate the submerged intentions of planned spaces. Yet this perspective seems misplaced in a context in which, following over two decades of Structural Adjustment, the state is now almost entirely absent. Here, resettlement townships now represent the material expression of a statist vision that has almost entirely evaporated. In this context, in which a totalising order is now manifestly absent, it seems unhelpful to imagine the practices through which people inhabit these spaces as 'subversive'. In clinging to a past vision of a planned urban modernity resettlers render its absence as problematic. To the extent they seek to redefine their relationships to the bureaucrats and planners they take to be responsible for this predicament, this is not by subverting their plans. Rather, broken promises are highlighted in calling for their realisation.

Conclusion

In this chapter, I have described the afterlife of a mid-century experiment in resettlement, highlighting how continued enchantment with a modernist experimental vision persists in disjunctive relation to an infrastructural reality perceived to lack these qualities. My discussion of the contemporary practices and understandings through which people now occupy these spaces makes evident how the project inflected ideas about urban life with utopian visions that continue to resonate. The 'failure' of the project is not understood as a matter of the experimental vision itself: it arises through juxtaposition of an anticipated modernity and a social and infrastructural reality seen as a negative instance of this modernist ideal.

The recent 'turn to ruins' helps to highlight these dynamics to the extent these accounts illuminate the complex imaginative, temporal and affective dynamics that arise through interactions with literally decaying materials. Helpful, also, is the insistence in recent work on the contextual specificity of these encounters. However, while the analytic lens of ruination opens up interpretive possibilities, the lens of the Volta Resettlement Project makes evident a limiting assumption inherent in much of this literature. In an influential paper, Dawdy (2010: 777) suggests that 'there is hope in ruins, in the suggestion that modernity can be surpassed'. Likewise, the celebration of ruination entails a critique of Cartesian planned space, and of a western metaphysics of presence. Edensor (2012: 844) suggests that ruins highlight how 'Modern attempts to cleanse, banish ambiguity, and order the memory of space are always disturbed by such disorderly spaces and by the ghosts they contain.' Gordillo (2014: 6) suggests processes of ruination reveal 'the critical power of negativity to disintegrate the positivity of the given, of things as they seem to be, and thereby to undermine any reified fantasy of a complete seamless whole'. In various ways, recent work celebrates the social, cultural and imaginatively creative possibilities that result from ruination: ruination is hopeful and productive as a counterpoint to the assumed hubris of modern concerns to order and discipline (Buchli 2013; Pelkmans 2013). Likewise, it is revelatory as the counterpoint to modernity's own hubristic understanding of fantasies of wholeness and completion. In line with broader critiques of modernist urbanism and planning, explicit or implied critiques of modernity eclipse what is ethnographically at stake in contexts where modernity exists as a condition that is absent but desired.

The ruins of this mid-century urban experiment are powerfully significant because of, not despite, the project's failure. These inspire a series of interlinked ideas about the gap between the possibility of the different, better, future anticipated by the experiment, and the actually existing circumstances that resulted from its failure. In some ways, these experiences resonate with accounts from other parts of the continent (Ferguson 1999; Larkin 2004), where modernity exists as a constitutive absence. Idealised visions of how life is elsewhere make apparent the deficit of life as lived here. However, the experimental context of the Volta Resettlement Project gives these dynamics a specific and arguably more pronounced form. As vanguards of what the nation could have become, the gap between post-independence promise and present reality is acute and intimate. Residents of resettlement townships encounter it daily in their interactions with now crumbling infrastructures that make apparent what they could have been and simultaneously what they are not. Possibility and problem, ideal and reality are mutually elicited through daily encounters with experimental remains.

Note

1 Letter from the chief planning officer for the attention of Mr Wright (Environmental Sanitation Division), EAK Kalitsi (resettlement officer) and George Nez (UN Mission to Accra), 25 May 1962.

References

Alexander, C. and Buchli, V. (2007). Introduction. In Alexander, C., Buchli, V. and C. Humphrey (eds), *Urban Life in Post-Soviet Asia*. London: UCL Press, 1–39.

Bonneuil, C. (2000). Development as experiment: science and state building in late colonial and postcolonial Africa, 1930–1970. *Osiris*, 15: 258–281.

Borup, M., Brown, N., Konrad, K. and van Lente, H. (2006). The sociology of expectations in science and technology. *Technology Analysis and Strategic Management*, 18: 285–298.

Buchli, V. (2013). *An Anthropology of Architecture*. London: Bloomsbury.

Butcher, D.A.P. (1970). The social survey. In Chambers, R. (ed.), *The Volta Experience*. London: Pall Mall Press, 78–102.

Chambers, R. (1970). Introduction. In Chambers, R. (ed.), *The Volta Experience*. London: Pall Mall Press.

Chambers, R. (1983). *Rural Development: Putting the Last First*. London: Longman.

Danby, M. (1970). House design. In Chambers, R. (ed.), *The Volta Experience*. London: Pall Mall Press, 164–178.

Dawdy, S.L. (2010). Anthropology and the ruins of modernity. *Current Anthropology*, 51: 761–793.

de Certeau, M. (1984). *The Practice of Everyday Life*. Berkeley: University of California Press.

Edensor, T. (2012). The ghosts of industrial ruins: ordering and disordering memory in excessive space. *Environment and Planning D: Society and Space*, 23: 829–849.

Ferguson, J. (1999). *Expectations of Modernity: Myths and Meanings of Urban Life in the Zambian CopperBelt*. Berkeley: University of California Press.

Ghanaian Times (1966). Editorial: our faith. *Ghanaian Times*, 24 January 1966, p.5.

Gordillo, G. (2014). *Rubble: The Afterlife of Destruction*. Durham, NC: Duke University Press.

Holston, J. (1989). *The Modernist City: An Anthropological Critique of Brasilia*. Chicago, IL: University of Chicago Press.

Johnson, A.A. (2013). Progress and its ruins: ghosts, migrants and the uncanny in Thailand. *Cultural Anthropology*, 28(2): 299–319.

Larkin, B. (2004). Degraded images, distorted sounds: Nigerian video and the infrastructure of piracy. *Public Culture*, 16: 289–314.

Lefebvre, H. (1991). *The Production of Space*. Oxford: Blackwell.

Mbembe, A. (2004). Aesthetics of superfluity. *Public Culture*, 16(3): 373–405.

Mensah, K.B. (1961). Problems of housing in connection with the Volta River Project: Resettlement and Compensation. Memorandum, 3 July 1961.

Miescher, S.F. (2012). Building the city of the future: visions and experiences of modernity in Ghana's Akosombo Township. *Journal of African History*, 53: 367–390.

Mitchell, T. (2002). *Rule of Experts: Egypt, Techno-Politics, Modernity*. Berkeley: University of California Press.

Nielsen, M. (2011). Inverse governmentality: the paradoxical production of peri-urban planning in Maputu, Mozambique. *Critique of Anthropology*, 31: 329–358.

Pelkmans, M. (2013). Ruins of hope in a Kyrgyz post-industrial wasteland. *Anthropology Today*, 29: 17–21.

Rabinow, P. (1989). *French Modern: Norms and Forms of the Social Environment*. Chicago, IL: University of Chicago Press.

Roy, S. (2007). Urban space, national time, and postcolonial difference: the steel towns of India. In Cinar, A. and Bender, T. (eds), *Urban Imaginaries: Locating the Modern City*. Minneapolis: University of Minnesota Press, 182–207.

Schwenkel, C. (2013). Post/socialist affect: ruination and reconstruction of the nation in urban Vietnam. *Cultural Anthropology*, 28: 252–277.

Shapiro, J.E. (2003). *Settling Refugees, Unsettling the Nation: Ghana's Volta River Resettlement Scheme and the Ambiguities of Development Planning, 1952-1970*. Ann Arbor, MI: University of Michigan.

Silver, J. and Marvin, S. (2015). *Powering Sub-Saharan Africa's Urban Revolution: An Energy Transitions Approach*. Working Paper, Department of Geography, University of Durham.

Stoler, A. L. (2008). Imperial debris: reflections on ruins and ruination. *Cultural Anthropology*, 23: 191–219.

Yarrow, T. (2011). Kinship and the core house: contested understandings of kinship and place in a Ghanaian resettlement township. In Edwards, J. and Petrovic-Steger, M. (eds), *Recombinant Knowledge: Or How Stratherian Concepts Travel*. Cambridge: Cambridge University Press.

16 The glorious failure of the experimental city

Cautionary tales from Arcosanti and Masdar City

James Evans, Gabriele Schliwa and Katherine Luke

Crises, contrasts and cautionary tales

It is hard to imagine two more archetypal experimental cities than Arcosanti, a commune-style urban laboratory founded in 1970 in Arizona, and Masdar City, a current-day living lab for clean-tech development in the Middle East. Both Arcosanti and Masdar City reflect experimental urban responses to moments of environmental crisis; the former originating in the environmental and oil crises that hit the United States in 1969 and 1973, and the latter gestating in the era of Al Gore's *An Inconvenient Truth* and the resource security concerns associated with the Iraq wars. But beyond that, the two places appear diametrically opposed. The contrast between Arcosanti, a product of the 1960s belief in progressive social forces, and Masdar City's contemporary neoliberal corporatism could not be sharper.

This chapter examines the underlying motivations and philosophy of these self-styled experimental cities to reveal continuities in the ways in which experimentation is articulated. The lived experience in each place suggests that neither has managed to embed its experiments into an urban culture or political system. But while the social and political aspects of urban experimentation are largely neglected, each place articulates a clear ideology of how urban society can be designed around new technologies. The current proliferation of urban experiments around the world reflects the same urgency to find alternatives to business as usual in the face of pressing global challenges and a similar ideological dependence on technology as the basis for a designed urban society. Arcosanti and Masdar City serve as cautionary tales for such approaches, highlighting the need to situate new technologies within a lived political and social context rather than designing cities around technologies.

Arcosanti – a laboratory for self-creation

Seventy miles north of Phoenix, Arizona, in the highlands of the Sonora Desert sits one of the most interesting settlements in North America. Founded in 1970 by Italian architect Paolo Soleri (1919–2013), Arcosanti is a self-styled urban laboratory for the discovery of new ways of human living (Soleri 1983). Auto-constructed by a constantly rotating group of some 7,000 volunteers over the last 40 years, Arcosanti enacts Soleri's principles of arcology – a philosophy which merges ideas from ecology into architecture (Soleri 1969). Arcology focuses on the creation of habitats for human beings that enable the progressive development of civilisation and intelligence through increasing complexity. The idea of miniaturisation plays a key role in this philosophy. Bringing people and things into closer proximity creates the conditions needed to facilitate greater levels of interaction. It is these interactions that generate complexity, liveliness and evolution. Arcology draws design principles from ecology, noting that higher life forms (such as bees, ants, apes), live in organised, dense settlements, while lower life forms (corals, moulds and so on) are spread out.

Soleri theorised that the shape of the city could change human behaviour and devised numerous 'arcologies', cities designed using arcological principles. In contrast to Broadacre City developed by Soleri's former teacher Frank Lloyd Wright and embodied in the two-dimensional sprawl of Phoenix, Arcosanti is a car-free, three-dimensional settlement. It is inspired by the walled cities of Italy and the pueblo dwellings of indigenous peoples in the southwestern United States, both places where Soleri felt society to be more vibrant and the environment better conserved. Arcosanti sits on 860 acres of land in total, much of which is rented for cattle ranching. Although the footprint of the city only covers 25 acres, Soleri would have said that the 860 acres are equally part of the urban form, as the small city footprint allows conservation of the surrounding desert.

Arcosanti was built in an extreme environment to show that sustainable living was possible anywhere and that the arcological model could be replicated. Soleri proposed hundreds of other (sometimes more outrageous) cities. Many, such as the Space Arcology, were never intended to be built but rather to push the limits of architecture and promote the densification of urban design. Density forms the main principle animating Soleri's urban designs, prefiguring subsequent planning orthodoxies concerning the desirability of compact cities and their role in facilitating creativity. Soleri terms the process by which cities generate complexity the 'urban effect'. While his designs have mostly been discussed in relation to their ability to address environmental resource constraints, the idea of the urban effect is based upon a set of arguments concerning its value in relation to quality of life and the enhancement of creativity (Soleri 2001).

Visually, Arcosanti is synonymous with its large half-dome structures, or apses, which ameliorate the extreme temperatures of the desert. Constructed from concrete, they have a high thermal mass, which, combined with the south-facing orientation, enables them to remain cool during the day and emit warmth at night. As Figure 16.1 shows, these domes are distinctive and almost otherworldly. The apses are based on the same earth cast construction principles as the Soleri windbells, which are produced by the foundries at Arcosanti and Cosanti, the original Soleri settlement, in the Paradise Valley suburb of Phoenix. Making

Figure 16.1 Arcosanti sign – the self-proclaimed urban laboratory, plus characteristic apse to right (source: authors).

bells generates income to support the broader Arcosanti project, although ironically they now enjoy at least as much fame as Arcosanti itself.

In the last 20 years, Arcosanti has focused on developing food production systems, using closed-loop design principles. Projects in greenhouses scattered throughout the settlement grow produce for a weekly farmers market to raise money for seeds, irrigation equipment, and greenhouse maintenance and any surplus is used in the communal kitchen. Arcosanti also had 60 acres under cultivation with corn and peaches. Soleri never supported these projects, which he thought distracted volunteers from the central goal of completing the city construction. After a flood on the Agua Fria River destroyed the dyke protecting the fields, the land was not replanted, and the orchard subsequently removed. The existing systems serve more to show how food production can be incorporated into an urban landscape, and, given the huge level of interest in sustainable food in the United States currently, serve to attract people to Arcosanti. Other efforts to develop a local economy, including a bakery and the Ferguson's free store (the only shop that ever filled space in the rooms devoted to commercial use in the residential East Crescent complex), similarly faltered given the lack of labour, funding and a local market. The total operating budget for 2012 was less than US$1 million (Tortorello 2012) and the daily cash economy is limited as the majority of residents earn only a minimum wage and many are simply volunteers that receive room and board in exchange for work.

Arcosanti is often portrayed as a typical eco-alternative community typified by self-sufficiency but in terms of its functions and services it was not intended to be that different from other towns. Arcology is based on the idea of self-containment rather than self-sufficiency, whereby 'leanness' tends towards increasing miniaturisation of form and function (Soleri 2002). Within the precepts of arcology, the primary motivation to reduce resource use is because this is the basis of progress towards higher levels of civilisation and well-being. The more concentrated a habitat is, the more interactions it produces between its component parts, which in turn produce more innovation, liveliness, and culture. Figure 16.2 shows examples of how this kind of density has been attempted at Arcosanti, with the settlement designed vertically as well as horizontally to intermingle work and living spaces. For Soleri, progress involves the internalisation of matter through miniaturisation and the growing complexity this affords. The concept of self-sufficiency is not just unrealistic but irrelevant and even anathema to this argument. Self-containment enables self-generation from within.

Within this understanding the primary role of the city becomes to function as a laboratory for self-creation. Arcosanti combines design and technological innovations to produce experimental living arrangements that address problems associated with land use, social integration, energy use, food production and transport (Grierson 2003). Soleri (2002) makes an explicit distinction between experimentation and the experiential process, whereby the laboratory becomes an ideal habitat for human development. Whilst the laboratory 'is necessary to develop an experiment and then take it to its breaking point … it is not the experiment that connotes learning but the laboratory' (Soleri 2001: 66). The goal of urban design is to create cities as laboratories to enable self-creation. From this perspective sustainability becomes part

Figure 16.2 Dense and three-dimensional urban form, mixing living and work spaces at Arcosanti. (Left to right): dorms, storefront space in the East Crescent, and the amphitheatre. (Source: authors.)

of a process of progressive self-transformation (Grierson 2003), which is clearly distinguished from adaptation per se. As Soleri states, 'we have to sell the Lean Alternative or we can sell out and adapt. *To adapt is to forego the very notion of the laboratory* [emphasis added]' (Soleri 2002: 24). Adaptation comes from without, whereas learning comes from within – 'the laboratory is the place where things that are uncommon and unproven are tested: a learning process by definition'.

The ideal of auto-construction, or self-building, flows from this thesis, as the city provides the medium for self-transformation. As Figure 16.3 shows, Arcosanti is permanently under construction – a fact that produces tensions within the community. Construction is necessary to maintain the city as one-third of their income comes from the tuition payments of workshop participants, who spend five weeks at Arcosanti learning the basics of arcology theory, earth casting, and concrete construction; but there is an increasingly limited budget for new projects. This has challenged their fidelity to Soleri's original plan. Everything adheres to a master plan (even the bells are based on his original casts and the foundry workers are not allowed to deviate from certain shapes and patterns). However, much of the design is now out of date or no longer feasible. Some projects have been abandoned, such as the energy apron, a seven-acre greenhouse designed to extend down the wall of the canyon above which Arcosanti sits and funnel hot air into the city for heating (Arcosanti 2015).

Arcosanti and the principles of arcology upon which it is based resonate with current debates concerning urban experimentation. Far from seeing urban experiments as something separate that can be grafted onto cities, urbanisation is cast as a process of self-creation whereby cities act as laboratories to enable learning. The city provides the socio-material context through which people and things are transformed into ever more tightly knit relationships, connecting people and things in ways that political and institutional accounts of urban change have tended to overlook (Bulkeley *et al.* 2014). The material fabric of the city becomes part of the process of self-invention and frames the process of urban learning.

Smart technology resonates with this way of thinking. Smart phones are the paragon of miniaturisation. They have internalised a huge amount of matter (address books, diaries, desktop computers, and so on) and facilitated unprecedented levels of communication and interaction. Smart grids are predicated upon reducing the distance that energy needs to travel by decentralising energy production and managing supply and demand on local levels. Smart districts achieve miniaturisation and densification through the use of ICT to increase connectivity. The modularity of arcological urban designs also allows them to be scaled to different sizes based on population, a concept most recently used in Soleri's designs for the Lean Linear City (Soleri *et al.* 2012), which is essentially a series of urban units that can be stitched together. These urban units are based upon the same kind of closed-loop thinking that animates urban sustainability systems from waste reuse to rainwater capture. However, for Soleri, efficiency gains are a side effect rather than aim of this process – the ultimate goal of the city is to act as the medium through which a more complex and hence innovative society can develop.

Figure 16.3 Construction projects underway in 2012 at Arcosanti (source: authors).

While Paolo Soleri passed away in 2013, the settlement continues under the management of the Cosanti Foundation, a not-for-profit educational organisation that aims to share Soleri's architectural concepts and philosophy with students, professionals and the general public, and ensure the continuing construction of Arcosanti. Initially intended to house 5,000 inhabitants, only 5 per cent of the originally planned settlement has been built and it struggles to attract more than 100 residents at any time. Soleri never wanted to build a community. He envisioned Arcosanti as a laboratory and worksite where workers lived for convenience, which should not be considered a permanent home until construction was complete. Many of the longest working employees do not live there, and neither did he for the last 20 or so years of his life. These intentions aside, community considerations have made the development of basic governance structures unavoidable. All new residents must complete the five-week workshop; after which, they are permitted to petition for residency, which is contingent upon finding a position (either paid or voluntary) within the settlement and approval by the Community Council. Even then, the new residents rarely live in the city itself but in 'cubes' at the bottom of the mesa.

Living in Arcosanti demands certain sacrifices. It is not possible to drive, there is no cable television, the closest grocery store is a 40-minute commute. And yet, people chose to live there and adapt their behaviour as such. It is not just about experimentation with architecture, but also styles of living and community building, with which not all of the issues have been successfully resolved. When Soleri lived at Arcosanti, he was the ultimate arbitrator of community complaints. Not only did the city adhere to his master plan but his specific rules also regulated residence, in many ways smacking 'of a puritan desert despotism' (Luke 1997: 171). As the founder of the project, he was widely respected and residents were willing to adhere to the rules in order to participate in the creation of an ideal city. The shared and progressive commitment to environmental sustainability also obscured the lack of democracy within the community. However, as Soleri slowly withdrew from leadership, numerous conflicts emerged over the attempt to establish a system of governance. A Community Council was created after Soleri moved back to Cosanti to manage day-to-day operations of living in the settlement but meetings were often dead-locked by small issues such as having dogs, quiet hours, and where people could smoke and drink. The older residents complained of disrespect and the younger residents felt unrepresented. In the summer of 2012, the residents of the cubes drafted a list of demands including more transparency and greater representation on the Community Council. Older residents resisted ceding control of the city to the transient population of younger volunteers passing through, in some cases driving them away from the settlement.

Questions loom as to the future of the city. There is an ongoing debate about shutting down construction and turning Arcosanti into a museum or a conference centre. Following the death of Soleri in 2013, the Cosanti Foundation solicited applications for the Strategic Plan Steering Committee comprised of 22 alumni and residents. The Committee will meet quarterly over the two-year planning process and has actively sought feedback from residents, alumni, and the Board of Directors. This is not just a social challenge but an economic one, as the ideological precepts upon which Arcosanti is based have created a splendid isolation from the broader industrial and commercial relations that nourish conventional urban growth (Luke 1997). Perhaps because of this, the arcology model never spread. While many have taken part in the workshops at Arcosanti, the impact of the project is unclear. Other arcologies have been proposed (Quick 2011), yet Arcosanti remains the only one ever built. Most of the current structures were completed in the 1970s and 1980s and it is hard not to conclude that it resembles what the city of the future might have looked like 30 years ago. Certainly the sweeping ideas of arcology, linking urban design to the self-realisation of the human species display an unfashionable type of generalising modernist thought.

That said, Arcosanti receives between 25,000 and 50,000 visitors per year, and on a visit in 2012, we were told that delegations of Chinese planners running into the hundreds were periodic visitors. Soleri also did some consulting in China and presented the Lean Linear City at the Beijing Center for the Arts in 2009 (Arcosanti 2009). Although the city is known globally, especially in Italy and Japan, within Arizona it is more commonly understood to be a commune than a comprehensive project about alternative urbanism.[1] That is also what it has become as it attracts both architecture students and other (mostly) young people often seeking a back-to-the-land, eco-village experience. However, Soleri especially was a very public figure. He was invited to design a footbridge in the posh neighbourhood of Scottsdale on the Arizona Canal and a highway overpass outside of Cordes Junction. His bells were even showcased at the 2015 Super Bowl in Glendale. Whether or not Arcosanti is viable or applicable to the design challenges of already existing cities is a question that Soleri never answered.

Masdar City – the test city

Fast forward 40 years and we find a second iconic urban laboratory that constitutes a full-scale constructed experimental city. In the Middle Eastern desert, about 17 km outside Abu Dhabi in the United Arab Emirates (UAE), Masdar City has been designed as the world's most ambitious eco-city project. Supporting Abu Dhabi's Economic Vision 2030 to transform from a natural resource-based to a knowledge-based economy, the city serves as a 'centralized test-bed for global renewable-energy and technology companies' (Masdar, 2016). As a high-density, pedestrian-friendly urban area it is designed to maximise convenience and reduce environmental impacts. Masdar City aims to be a zero-carbon, zero-waste global benchmark for sustainable development, intended to become the Silicon Valley of the energy sector (Masdar 2014a). This claim is subject to continual reworking and repetition. Comparing what California is to the microchip, Mohammed El Ramahi, Masdar's head of utilities and asset management, says Masdar City can also be the Silicon Valley of water (Henley 2013). In about six square kilometres of special economic zone, 50,000 residents and hundreds of clean-tech industries are supposed to find their place.

In contrast to Arcosanti, Masdar, which is the Arabic word for 'source', aims to connect to the global economy to make Abu Dhabi the leading provider of renewable energy knowledge and technology. The Masdar Company, a subsidiary of the

Abu Dhabi Government-owned Mubadala Development Company, was established in 2006 as a catalyst for the Emirate's economic diversification, with the mission to advance renewable energy and sustainable technologies through education, research and development, investment, commercialisation and adoption (Masdar 2013). The five business units – Masdar City, Masdar Capital, Masdar Clean Energy, Special Projects, and the Free Zone – are complemented by the Masdar Institute, a research-driven graduate university developed in cooperation with the Massachusetts Institute of Technology (MIT). The university is considered to be the key source of R&D and aims to become one of the world's leading academic institutions in the field of energy technology. Its strategic location within Masdar City and the quality of students and education are considered crucial success factors (Masdar 2013).

Beginning in 2010, the Masdar Institute and its 330 students were the first tenants of Masdar City. During our walkthrough, tour guide Mohammed Al Hosany emphasised that the design and the equipment of the campus would enable students to focus purely on their studies, creating a mutually beneficial relationship between the urban space and its tenants and effectively integrating the business units Masdar City and Masdar Institute. Currently, Masdar City's buildings reduce energy demand by 56 per cent and potable water demand by 54 per cent compared to traditional Middle Eastern structures. Renewable energy sources provide 100 per cent of local energy demand through on-site electricity generation by a ten megawatt solar photovoltaic plant and a one megawatt rooftop solar photovoltaic installation. The street layouts are designed to mitigate the effects of the hot summer sun, as walkways and plazas are oriented away from the south (Figure 16.4). These features enable the cityscape to feel up to 20°C cooler than a conventional urban area in Abu Dhabi and further reduce energy demand (Albanese 2013).

Our visit in 2013 revealed a city under construction and left the visitor with the sentiment of a place typifying the 'malls without walls' style of urban development projects, such as Business Improvement Districts, that are partly or entirely

Figure 16.4 Masdar's architectural design directs desert sun and winds to enhance urban living. (Left to right): school class in front of the library, wind breakers and Masdar Cooling Wind Tower. (Source: authors.)

Figure 16.5 Construction works at Masdar City in 2013 (source: authors).

financed by private capital to create a consumer-friendly environment (Graham and Marvin 2001; Minton 2006). Masdar's homepage offers licensing packages for companies interested in immigrating to the Free Zone but no information on residential living. Masdar Free Zone represents a special economic zone where new economies can be trialled in order to attract green industries through unrestricted flows of capital such as zero percent import taxes, no currency restrictions and other financial incentives (Caprotti 2014). The Q&A for the Masdar Free Zone explains that the residence visa for company staff is valid for a period of three years from the date of issue and family members can accompany them if eligible per the immigration rules. Much like Arcosanti, Masdar faces the difficult task of attracting residents to an extreme desert environment lacking even the cultural and social amenities of neighbouring Abu Dhabi City (Figure 16.5). In 2014, it was announced that Masdar is getting its first 500 houses, seven years after construction began, and far removed from its original goal to host over 1,500 businesses and 50,000 residents in 2,000 houses by 2016.

While representing the 'world's largest cluster of high-performance buildings that together create a real-time laboratory to monitor and study how cities use, conserve, and share resources' (Masdar 2014a), masters students on site report that real cases from Masdar City are partly but not necessarily included in their education. Reporting from first impressions how it feels to actually live in Masdar City, a student writes on her blog:

> I keep telling people that it feels like I'm living in a psychology experiment. Every time I flip a light switch in the living room and the faucet in the bathroom starts running, or I desperately push all buttons on the stove to try to turn on a burner, I can't help looking over my shoulder and wondering if there's a scientist observing my behavior and reactions in this strange environment.
>
> (Rants and Rambles Blog 2010)

Looking at recent reviews, not much seems to have changed. Tourists as well as locals have mixed feelings and report on the one hand about being amazed by the futuristic architecture and on the other finding a strange atmosphere on site (Ouroussoff 2010; Kingsley 2013). Generally, visitors struggle to spend more than one interesting hour in Masdar and the personal rapid transit system is invariably the highlight (TripAdvisor 2014). Initially a high-profile system to provide personalised public mass transport throughout the city with some 80 docking stations and thousands of vehicles, it has been one of the declared failures. As Figure 16.6 shows, the personal rapid transit system was intended to enable car-free transport in the lower level, which turned out to be problematic. After the financial crisis in 2008, developers had to cut their ambition of a city-wide roll out of the vehicle that seems like something from the *TRON: Legacy* movie, partly because there was no way of persuading building developers to share the costs of the personal rapid transit stations (PRT Consulting 2010). The Masdar City website, however, still announces the personal rapid transit that connects the car park with the Masdar Institute to be an attraction for visitors.

Looking more closely into how Masdar City is conceptualised within the broader philosophy of the company raises questions over whether this city will ever become a place for human living or if the social dimension within the company's 'holistic approach' has been marginalised from the beginning. In the shadow of extensive marketing and high-level policy documents, society is generally an afterthought in the master plans of cities like Masdar (Caprotti 2014). In contrast to Arcosanti's 'auto-construction' based on volunteers, Masdar's construction process itself is typically described in passive terms and the lack of attention to labour conditions is not surprising. Typically, construction in Abu Dhabi is done by low-skilled migrant workers who live in low-rent areas close by until they need to move on to the next mega project (Crot 2013; Caprotti 2014). To counter this, compliance with national standards on health and safety is being advertised; for example, through Masdar's campaign for workers' heat stress awareness (Masdar 2014b). Politically, the lack of democratic structures in Masdar echoes the situation at Arcosanti, although in this case is also nested within an opaque development corporation and a largely non-democratic national political structure.

Masdar City's key performance indicators emphasise customer satisfaction to measure social aspects of sustainable development (Cugurullo 2013). Staff and

Figure 16.6 Personal rapid transit in Masdar City's undercroft. Instead of 80 docking stations spread underneath the city's surface as originally planned, only one station connects the car park with the Masdar Institute. (Source: authors.)

corporate material emphasise the project's superlative role as a living lab for corporate testing, stating that 'Masdar City is an emerging global clean technology cluster that places its resident companies in the heart of the global renewable energy and clean-tech industry' (Masdar 2014c). The role of the city is to function as a research environment to attract these industries, with Masdar City presented as a set of buildings and infrastructures rather than a coherent urban system. For example, Masdar City's urban living lab offer is essentially composed of built infrastructure, corporate clustering, a special economic zone and university research capacity. The sales pitch is to companies rather than people and locating in Masdar City gets them 'access to this living laboratory … [and] to other companies that they may want to have a relationship with, and very bright students and faculty who have a relationship with one of the world's best universities' (Fearn 2009). Fairness towards new customers makes the city 'socially acceptable' from a developer point of view whilst social cohesion is reached merely at the symbolic level without being part of its core activities and policies (Cugurullo Chapter 14, this volume).

Masdar City's main internal driver is economic diversification as an adaptive measure for a country built on oil in a world that increasingly acknowledges the end of the fossil fuel era. In contrast to Arcosanti's image, Masdar City is often referred to as a neoliberal solution to contemporary urban problems in the media and academic circles (Hodson and Marvin 2010; Kaika and Swyngedouw 2012). The leadership of the eco-urbanism movement is strongly focused around particular corporate and governmental interests. Masdar City's property developer states: 'We want Masdar City to be profitable, not just sunk cost. If it is not profitable as a real-estate development, it is not sustainable' (quoted in Bullis 2009: 2). As such, the objective of Masdar City is to turn the urban development process into a set of products that can be combined and deployed in any urban context. While Arcosanti and Masdar appear to be political opposites, there are structural similarities between their logics of upscaling, which are based on developing and demonstrating modular urban solutions that promise a high degree of portability. This is a key comparison between the two cities that reflects a design-led vision to create replicable solutions.

Despite its reputation as an expensive experiment in urban sustainability, Masdar City does not seem to perceive the entire city as an experiment. Rather, like Arcosanti, the city forms a laboratory within which experiments can be staged. Technology scientists as well as clean-tech companies attempt to learn from the experiments conducted within the city with respect to certain sectors, such as architectural design, renewable energy technologies deployed or materials used (Sgouridis and Kennedy 2010; Janajreh *et al.* 2013). Nawal Al-Hosany, the director of sustainability at Masdar City, refers to experiments within the city: 'As we build the city we're constantly experimenting with a lot of new technology, and we're constantly learning from the process' (quoted in Dumaine 2013). The city does not seem to constitute the subject of learning but rather exists as a passive infrastructure in which products can be tested. This indicates the complexities and implications of how the boundaries of the experiment are drawn in relation to the city. Just as Cugurullo (2013) analysed Masdar City's three dimensions of sustainability as 'Inside the Core', 'Around the Core' and on 'The Surface', different layers of experimentation can be found within the city, around it and lying on its surface. Inside the core, experimental boundaries come down to testing and showcasing discrete clean-tech innovations. Around Masdar City, social science researchers as well as residents can sense, observe and study the effect the instrumentalisation of a city has for social sustainability and cohesion. On the very surface, Masdar City became an eco-city prior to its birth in public discourse that predated any form of evaluation and assessment.

Although branded as a city, Masdar City is more accurately described as a clean-tech cluster. Clusters are commonly described as a formation of interconnected businesses located in close geographical proximity, allowing them to share key resources, build up trust through demonstration and communication, and develop commercial advantages in efficiency, effectiveness and the rate of innovation (Porter 1998). It is expected that by 2020, the leading clean-tech clusters will be found not only in traditional innovation areas, like Clean-tech Valley near San Francisco, but increasingly in zones where flexible governance and rapidly applied, clean-tech focused economic policies are deployed, such as Masdar City or Copenhagen (Gray and Caprotti 2011). What determines the success or failure of a cluster is if it allows an entrepreneurial eco-system for clean energy, water and sustainable living to develop and thrive and, as such, be able to stand on its own financially, making it replicable in places where public investment is not available (Zuberi 2008). With US$22 billion in governmental investment, Masdar City itself is probably too big to fail. Nevertheless, questions remain concerning whether or not it can produce replicable and functional advancements in sustainable design given that human scale interaction is not at its core.

Design displaces politics

Walter Benjamin's modest take on the future suggests that, 'everything will be as it is now, just a little different' (Agamben 1993: 44). Urban experiments need an umbilical cord connecting them with existing socio-economic systems and an operational governance system to involve and represent residents if they are to be viable. The extreme alternatives manifested by Arcosanti and Masdar are self-defeating because they are disconnected from existing urban political systems, making them neither adaptable nor adoptable. Being geographically isolated means these experiments are stuck nowhere, doomed to be glorious failures. In 40 years, Masdar City will be the Arcosanti of our times – an unfinished monumental oddity to the unswerving belief in the ability of global corporate capitalism to deliver change that held sway in the early twenty-first century.

Neither place has any trouble drawing visitors but both struggle to attract and retain residents, which is problematic for any city. Urban transformation requires social, cultural, physical and technological systems to be embedded in the fabric of daily life, and experimentation needs to involve all these dimensions to understand what will work in practice. Even more so than in Arcosanti, Masdar

City has so far failed to embed its experiments into an urban culture or social infrastructure. Physical density is not enough to produce useful innovations. This social deficit is compounded by the geographical distance of both places from other vibrant urban centres, which makes them less appealing for residents to stay beyond their work or studies.

This tension expresses itself socially. In both places, residents essentially function as test pilots rather than communities. Characterised by the production of design experiments that neglect the specific social and broader political dimensions of the city, the 'social' or 'living' part of these cities has been overlooked by the designers and has evolved in an accidental way, punctuated by occasional eruptions. Their limited ability to function as communities prevents them from realising their ultimate political vision to drive transformative urban change. Soleri never imagined Arcosanti to be an eco-village, commune, or more permanent community, yet there are many long-term residents anxious as to the stability and security of the settlement since Soleri's passing. Masdar City similarly lacks social and political glue to hold its residents together; a gated dormitory community that is experienced as a non-place by visitors. Both places are design experiments in which the details of governance, whether traditional or experimental, are overlooked. Innovation displaces governance with the result that the design of urban systems stands in for politics.

In terms of their physical structures both Arcosanti and Masdar City rely on compact urban layouts achieved through three-dimensional planning and eco-sensitive design in relatively harsh desert environments (it is striking that the sun-shielding structures in both Figure 16.1 and Figure 16.5 are based on half-dome shapes). Here, design plays another role; the attraction of like-minded actors, whether they be clean-tech corporations inspired by the new university and high-tech transit systems or sustainable architecture enthusiasts and professionals interested in high-density living. Experimentation reflects a broader neoliberal obsession with invention, whether of the self or of products, and, in elevating processes of experimentation, both Arcosanti and Masdar privilege certain people and organisations. While it is unfair to critique the designers for this as they never focused on these places as lived environments, this does indicate a very limited view of what comprises an experimental city. These are 'uncivic cities' that almost completely neglect the social and political aspects of urban experimentation. This kind of instrumental experimentation produces sterile technological interventions that are unlikely to take hold.

This is a salutary point in relation to the current trend to turn cities and districts into platforms for experimentation that reframe policy and decision-making around innovation (Bulkeley and Castán Broto 2012). For funding bodies and governments, urban experimentation offers a way to cajole intransigent and risk-averse authorities into generating and adopting innovative urban solutions without having to engage in highly politicised structural or policy reform. This modular approach, based on discrete urban experiments, promises portability but produces physically fragmented and politically deficient cities. Current-day approaches like urban living labs emphasise a more democratic and user-centric approach, but still struggle to place the social, political and cultural dimensions of the innovation agenda on the same footing as technology (Evans and Karvonen 2011, 2014). The current proliferation of urban living labs, especially across Europe and Asia, heralds a design-led urbanism united in the belief that if the process is designed correctly, then the outcome will be favourable. This risks being too socially and politically limited to seed widespread and durable change (Karvonen and van Heur 2014).

In Arcosanti and Masdar City, the more interesting learning is occurring in ways that were not intended by the designers: through organic political and social struggles. The emergent properties of each experiment are what is interesting, rather than what was intended. Paolo Soleri's challenge to create better urban habitats in which to experiment remains worthy and reverberates through a broader set of agendas and efforts to recuperate the city and urban form as something more lively, responsive and, of course, sustainable. But in taking up this challenge it is vital not to let design displace politics or to risk resigning ourselves to experiments that will remain forever frozen in place.

Note

1 On the role of communes in the Southwest, see Auther and Lerner 2011.

References

Agamben, G. (1993). *A Comunidade Que Vem*. Lisbon: Presença.

Albanese, M. (2013). *How She Leads: Nawal Al-Hosany, Masdar*. [Online]. Available: http://www.greenbiz.com/blog/2013/06/10/how-she-leads-nawal-al-hosany-masdar [25 April 2015].

Arcosanti. (2009). *Exhibition – BCA China – Preparation*. [Online]. Available: https://arcosanti.org/node/8074 [11 April 2015].

Arcosanti (2015). *Strategic Planning Progress Report*. [Online]. Available: http://arcosanti.org/node/13995 [19 April 2015).

Auther, E. and Lerner, A. (eds) (2011). *West of Center: Art and the Counterculture Experiment in America, 1965–1977*. Minneapolis: University of Minnesota Press.

Bulkeley, H. and Castán Broto, V. (2012). Government by experiment? Global cities and the governing of climate change. *Transactions of the Institute of British Geographers*, 38: 361–375.

Bulkeley, H., Castán Broto, V. and Maassen, A. (2014). Low-carbon transitions and the reconfiguration of urban infrastructure. *Urban Studies*, 51: 1471–1486.

Bullis, K. (2009). *A Zero-Emissions City in the Desert*. [Online]. Available: http://www.technologyreview.com/featuredstory/412195/a-zero-emissions-city-in-the-desert/ [12 January 2015].

Caprotti, F. (2014). Eco-urbanism and the eco-city, or, denying the right to the city? *Antipode*, 46: 1285–1303.

Crot, L. (2013). Planning for sustainability in non-democratic polities: the case of Masdar City. *Urban Studies*, 50: 2809–2825.

Cugurullo, F. (2013). How to build a sandcastle: an analysis of the genesis and Development of Masdar City. *Journal of Urban Technology*, 20: 23–37.

Dumaine, B. (2013). *In Abu Dhabi's Energy Oasis, Setbacks and Progress.* [Online]. Available: http://fortune.com/2013/04/29/in-abu-dhabis-energy-oasis-setbacks-and-progress/ [12 January 2015].

Evans, J. and Karvonen, A. (2011). Living laboratories for sustainability: exploring the politics and epistemology of urban adaptation. In Bulkeley, H., Castán Broto, V., Hodson, M. and Marvin, S. (eds), *Cities and Low Carbon Transitions.* London: Routledge, 126–141.

Evans, J. and Karvonen, A. (2014). 'Give me a laboratory and I will lower your carbon footprint!' – Urban laboratories and the pursuit of low carbon futures. *International Journal of Urban and Regional Research*, 38: 413–430.

Fearn, H. (2009). *The Green University at the Heart of a Living Lab.* [Online]. Available: http://www.timeshighereducation.co.uk/news/the-green-university-at-the-heart-of-a-living-lab/408092.article [16 January 2015].

Graham, S. and Marvin, S. (2001). *Splintering Urbanism: Networked Infrastructures, Technological Mobilities and the Urban Condition.* London: Routledge.

Gray, M. and Caprotti, F. (2011). Cleantech clusters and the promotion of the low carbon transition: criteria for success and evidence from Copenhagen, Masdar and online platforms. *Carbon Management*, 2: 529–538.

Grierson, D. (2003). Arcology and Arcosanti: towards a sustainable built environment. *Electronic Green Journal*, 1: 1-18.

Henley, W. (2013). *The Countries Vying to be the Centres of Excellence for Water Technology.* [Online]. Available: http://www.theguardian.com/sustainable-business/new-centres-excellence-water-technology-innovation [19 January 2015].

Hodson, M. and Marvin, S. (2010). Urbanism in the anthropocene: ecological urbanism or premium ecological enclaves? *City*, 14: 298–313.

Janajreh, I., Su, L. and Alan, F. (2013). Wind energy assessment: Masdar City case study. *Renewable Energy*, 52: 8–15.

Kaika, M. and Swyngedouw, E. (2012). Cities, natures and the political imaginary. *Architectural Design*, 82: 22–27.

Karvonen, A. and van Heur, B. (2014), Urban laboratories: experiments in reworking cities. *International Journal of Urban and Regional Research*, 38: 379–392.

Kingsley, P. (2013). Masdar: the shifting goalposts of Abu Dhabi's ambitious eco-city. *WIRED*, 17 December 2013. [Online]. Available: http://www.wired.co.uk/magazine/archive/2013/12/features/reality-hits-masdar/viewgallery/330478 [1 January 2015].

Luke, T. (1997). Developing an Arcological politics: Paolo Soleri on ecology, architecture and society. In Luke, T. (ed.), *Ecocritique: Contesting the Politics of Nature.* Minneapolis: University of Minnesota Press, 153–176.

Masdar (2013). *Masdar Corporate Brochure.* [Online]. Available: http://www.masdar.ae/en/media/detail/masdars-corporate-brochure [17 January 2015].

Masdar (2014a). *Masdar City at a Glance.* [Online]. Available: http://www.masdar.ae/en/masdar-city/detail/masdar-city-at-a-glance [17 January 2015].

Masdar (2014b). *Masdar Campaigns for Workers' Heat Stress Awareness.* [Online]. Available: http://www.masdar.ae/en/energy/detail/masdar-campaigns-for-workers-heat-stress-awareness [17 January 2015].

Masdar (2014c). *Masdar City Free Zone.* [Online]. Available: http://www.masdar.ae/assets/downloads/content/272/masdar_city_free_zone_brochure_eng.pdf [17 January 2015].

Masdar (2016). *Sustainability.* [Online]. Available: http://www.masdar.ae/en/masdar-city/detail/sustainability [28 January 2016].

Minton, A. (2006). *The Privatisation of Public Space.* London: RICS.

Ouroussoff, N. (2010). In Arabian Desert, a sustainable city rises. *The New York Times*, 25 September 2010. [Online]. Available: http://www.nytimes.com/2010/09/26/arts/design/26masdar.html?pagewanted=all [3 January 2015].

Porter, M. (1998). Clusters and the new economics of competition. *Harvard Business Review*, 76: 77–90.

PRT Consulting (2010). *Why Has Masdar Personal Rapid Transit (PRT) Been Scaled Back?* [Online]. Available: http://www.prtconsulting.com/blog/index.php/2010/10/16/why-has-masdar-personal-rapid-transit-prt-been-scaled-back/ [3 January 2015].

Quick, D. (2011). Floating megacity designed for the banks of the Mississippi. *Gizmag*, 3 February 2011. [Online]. Available: http://www.gizmag.com/pyramid-arcology-designed-mississippi/17787/ [17 January 2015].

Rants and Rambles Blog (2010). *Rants and Rambles: I Live in a Spaceship in the Middle of the Desert.* [Online]. Available: http://squidskin.blogspot.co.uk/2010/09/i-live-in-spaceship-in-middle-of-desert.html [3 January 2015].

Sgouridis, S. and Kennedy, S. (2010). Tangible and fungible energy: hybrid energy market and currency system for total energy management. A Masdar City case study. *Energy Policy*, 38: 1749–1758.

Soleri, P. (1969). *Arcology: The City in the Image of Man.* Cambridge, MA: MIT Press.

Soleri, P. (1983). *Arcosanti: An Urban Laboratory.* San Diego, CA: Avant Books.

Soleri, P. (2001). *The Urban Ideal.* Berkeley, CA: Berkeley Hill Books.

Soleri, P. (2002). *What if? Quaderni.* Paradise Valley, AZ: Cosanti Press.

Soleri, P., Kim, Y., Anderson, C., Nordfors, A., Riley, S. and Tamura, T. (2012). *Lean Linear City: Arterial Arcology.* Paradise Valley, AZ: Cosanti Press.

Tortorello, M. (2012). An Early Eco-City Faces the Future. *New York Times*, 15 February 2012. [Online]. Available at: http://www.nytimes.com/2012/02/16/garden/an-early-eco-city-faces-the-future.html?_r=0 [19 April 2015].

TripAdvisor (2014). *Masdar City Reviews – Abu Dhabi, Emirate of Abu Dhabi Attractions.* [Online]. Available: http://www.tripadvisor.co.uk/Attraction_Review-g294013-d2409120-Reviews-Masdar_City-Abu_Dhabi_Emirate_of_Abu_Dhabi.html [19 January 2015].

Zuberi, B. (2008). *Masdar City: Not a Showcase, but an 'Entrepreneurial Eco-system'.* [Online]. Available: http://bilalzuberi.com/2008/04/10/masdar-city-not-a-showcase-but-an-entrepreneurial-eco-system/ [19 January 2015].

17 Post carbon cities

Distributed and decentralized and demodernized?

Stephanie Pincetl

Introduction

The idea of post carbon cities has romantic appeal. Quiet, clean, and green with people moving around on public transportation, walking or biking purposefully to work or play in ways that have evolved to dispense of fossil fuels – a throwback to Ecotopia (Callenbach 1975). And each city reflecting its bioregional constraints and potentials (Register 2002). But beyond the previous imaginings of what such cities might look like, post carbon cities call into question what cities are for. Current modernist cities are the 'workshops' of the world economy (Storper 2013), places of production and consumption, supported by centralized infrastructures of power, water, sanitation and distribution systems, made possible by the extraordinary versatility and power of fossil fuels (McNeill 2000; Smil 2008). Fossil fuel energy is synchronous with the modernist city and global capitalism. While all cities have functioned as centres of power, trade, and/or religion, *modern* cities are truly about economic activity. The success of cities is predicated on whether they sustain economic performance or else they decline. People's well-being is judged by wages and jobs in cities. But is the post carbon city just that same city with the same drivers, minus fossil energy as the fuel driving the economy? Or is the transition something deeper and epochal?

Old European and Asian cities, built with human labour, are dense and human scale. Fossil fuel cities spread out; they can be suburban and/or tall and dense, serviced by infrastructures never imagined before. They harbour a concentration of people that is unprecedented in human history. As is well known, not only will this once-in-300-million-years energy supply begin to taper out, but its use is profoundly disrupting global climate and biogeochemical patterns. Debates rage about peak oil supply versus peak oil demand and these may never be resolved until some type of collapse occurs. Given the climate crisis, they may not even be worthy of further discussion. Rather, a transition away from fossil fuels is what is required. And while there is a great deal of analysis that shows fossil fuels can be successfully substituted and that deep decarbonization is possible (Williams *et al.* 2014), the transformation in energy supply itself implies profound change in urban form and governance, infrastructure and infrastructure governance, and a cascade of changes at multiple scales, including the global economy and its regulation, as well as the ways in which nearly everything is done.

To transform existing cities into post carbon ones, current urbanization patterns and forms will have to change to accommodate new technologies, the harvesting of local resources, and more localized production and consumption. Further, the gradual depletion of fossil energy will be accompanied by increasing scarcity of readily accessible, high-grade minerals and ores (Kaufman and Cleveland 2007). This will also necessitate changes in manufacturing, infrastructure, communications technologies, and so on – the list is long. There will be greater reuse of materials, materials mining and repurposing.

The post carbon city will be distinct from the pre-industrial city, using new scientific understandings of ecosystems and hydrology, better and more efficient technologies to enable novel blue–grey–green infrastructures and energy supplies. Cities will look and function differently, and, by necessity, they will challenge current governance structures, rules, regulations, norms and fundamentally, western notions of private property, individual 'rights' and land uses. The transition is not only about urban form and infrastructure but most importantly and fundamentally will require a rethinking of the role of cities and how we inhabit them. For these transitions to take place, experimentation must be allowed to occur. Risks need to be taken. One avenue is that of Jaime Lerner's 'Urban Acupuncture', precise interventions in specific locations (Lerner 2014). These changes – one experiment at a time – provide examples that can be built on. Often, however, they remain successful experiments, but replication or scaling up is not done for many reasons. So the challenge is to create experiments that change path dependencies or lock-in. One could say that the current drought in California is forcing experimentation that may ultimately change patterns profoundly. Cities and water agencies are re-incentivizing turf replacement and the state is considering a law that forces change in landscaping away from turf. This is experimentation on a large scale. It will have the effect of reducing carbon emissions because less water will need to be pumped and transferred.

But more profoundly, it is likely that a post carbon city will necessitate a new economy that is not predicated on economic growth and consumption where efficiency is a driving metric. Modernism and the modernist city are organized to deliver efficient services. Capitalist economic growth is predicated on greater and greater efficiencies in production and the expansion of markets, each of which requires ever more sophisticated techniques and technologies, from the autonomously driven car to more and more efficient use of each element of an agricultural product, and the invention of the 'next best thing'. Perhaps one can point to economies that are stagnant for natural experiments as they are not increasing their fossil fuel consumption but these would be sad and reactive experiments, not proactive and intentional ones.

The sustainable city

Definitions of sustainable cities mirror the Brundtland Commission report (1987), defining sustainability as involving environment, economy and equity. Sustainable city initiatives and discussions reference the need to maintain healthy economic conditions while protecting the environment and improving social equity. Sustainable city science and practice is underpinned by the belief that cities can become more sustainable by the reduction of inputs, automatically translating into the mitigation of environmental impacts, but has not generally engaged with the issue of carbon dependence nor the function of cities (Portney 2003; Mol *et al.* 2009). Much of the work has been measurement-oriented – quantifying urban environmental footprints (Wackernagel and Rees 1996; Folke *et al.* 1998), formulaic (the IPAT equation), and indicator-driven (McGranahan and Satterthwaite 2003, among many others). Even more specifically, industrial ecology, life cycle analysis, material flows and calculations of resource productivity of products and services (Ritthoff *et al.* 2002) have contributed new insights and understandings on the appropriation and transformation of nature involved in sustaining cities. A tremendous amount of quantitative research has been generated in the engineering and environmental sciences in this regard. There has been a standardization of models and indicators, such as Life Cycle Analysis by the International Organization for Standardization, along with the resurgence of urban metabolism analysis with statistical profiling of cities (Pincetl *et al.* 2014). New models purporting to quantify urban ecosystem services, such as trees, wetlands and green roofs, have been developed, including their hedonic value (Kong *et al.* 2007; Pataki *et al.* 2011; Pincetl 2012). The application of economic valuation has been used to argue for the value of urban nature (Garrod and Willis 1999), although this approach has done little to truly convince policy makers that the cost/benefit of ecosystem services is broadly worthwhile, and unlike sewage sanitation systems, they remain seen as costs that have squishy benefits. Surely one aspect of the difficulty is that 'ecosystem services' are difficult to value in contrast to the ways in which traditional hard infrastructure is. They require new knowledge, new management systems and new financing, and their individual value (e.g. a tree) is largely dismissible in terms of ecosystem services. Though often argued for in terms of green jobs creation, ecosystem services struggle for legitimacy compared to hard grey infrastructure and suffer from lack of good science and professionalism. At the same time, Ibanez and Katsikis (2014) point out, these approaches remain largely descriptive, quantitative questions of performance and efficiency, prioritizing energy, material and climatic optimization. Sustainability is doing better with the same – implementing economized bioswales instead of building stormwater pipes.

The National Research Council (2013) in a workshop on urban sustainability, cast sustainability as a tool for long-term growth, enhancing economic competitiveness through economic revitalization for cities experiencing declining economies, and as a method to support neighbourhoods among other goals, including reducing environmental impacts. Others, including Kates (2010) argue that to move urban areas toward sustainability, new urban forms, infrastructure, technologies and environmental management are needed and new behaviours, institutions and policies are required to innovate and implement them. This requires bringing together the science and technology of habitability, efficiency and environment with the practice of planning, building and financing cities.

Underlying urban sustainability science is the acceptance and presumption that humans will primarily be urban dwellers going forward. There is an urgent need for cities to change so they pose less of an environmental burden on the planet and at the same time ensure habitable environments for people. Thus, in many ways sustainable cities are ones that cause less impact on the environment due to substantive changes in efficiencies of technologies and new technologies, as well as form and management. But their function and governance remains unchanged: to be workshops for the global economy, their inhabitants labour in that scheme. Sustainability programmes have been weak in their ability to accomplish the third leg of the stool, improving equity, as they are in a deep sense 'green' technocratic fixes aimed to maintain economies as they exist today. Is a post carbon city a sustainable city?

If the hydrocarbon revolution is synchronous with capitalism and urbanization, is a post carbon city simply a city as we know it, but with alternative sources of energy? Or does post carbon lead to fundamental and dramatic change that cascades throughout the economic and socio-technical systems that support current human populations? And what then does that mean for capitalism and growth as the operant principles of human activity of the past century and a half, and the central raison d'être of cities? For if indeed carbon is the lubricant of modernity, which has included capitalism, industrialization, urbanization and accelerated economic growth, then without it things change dramatically. Perhaps there is a low carbon transition that enables a soft landing and a slower shift (Williams *et al.* 2014) but this too will require a massive change in economic planning and management, as well as governmental systems, probably little different than what would be required for a post carbon future. Sustainable cities today are imagined as integrating more ecosystem services such as stormwater bioswales that are expected to reduce stormwater waste, increasing tree canopy cover in the unsubstantiated aim of reducing air pollution impacts and sequestering carbon. There are now targets developed for desirable tree canopy cover in cities in the United States developed by the US Forest Service along with tree environmental improvement calculators, and recommendations about stormwater capture promulgated by the US Environmental Protection Agency. Mitigation of current conditions is the sustainability path put forward here.

Whether post carbon or low carbon, I would argue that a sustainable city and a post carbon city are not the same. A low or post carbon urbanization will look and feel deeply different and will require substantial changes in social organization and governance. It will involve deeply different relationships with near-by and far-flung hinterlands, a rediscovery of the commons, new property rights regimes and redefinitions of rights and obligations. It will require degrowth (in the wealthy west), contraction and new values. Degrowth does not mean doing less of the same, it means different activities, forms and uses of energy, different relations, allocations of time between paid and non-paid work, and different relations with the non-human world (Kallis *et al.* 2015).

The concept of sustainability emerged from a historic moment of the collapse of communism, the rise of neoliberalism and the spread of state-centred kleptocratic capitalism in China and Russia. It was an attempt to balance economic growth and development with management of environmental resources such that they would not be degraded to the point of not being able to provide subsequent generations with resources, with cobbled together language about equity. Sustainable city paradigms retain the goals of efficiency, while post carbon cities may not need such organizing principles as time will be less compressed. Today, with the assistance of fossil energy, tasks can be done rapidly. With less fossil energy available and energy sources that may be more climate or weather dependent, it maybe that things will take more time. Canning food, for example, rather than going to the grocery store to buy vegetables out of season that are imported from far-flung places.

Sustainability was a means to patch up the deep differences between north and south, and to nudge a more responsible capitalism to provide this balance. Only it has not worked. It ushered in the Kyoto Protocol and the further privatization of the public sphere with the ideology of selling carbon credits. Naomi Klein (2014) notes the timing of scientists calling for radical cuts in greenhouse gas emissions in 1988 and the signing of the world's largest bilateral trade agreement between Canada and the United States followed by the North American Free Trade Agreement unleashing the liberalization of world markets fuelled by unprecedented burning of fossil fuels and materials extraction to support expanding trade.

Rethinking the role of cities

In the past 150 years, the role of cities has increasingly become to serve as the economic engines for the global economy, obscuring other important reasons why people are attracted to cities and why cities are important for humans. Marx wrote of the isolation of rural life and its apolitical privatized apartness (Draper 1978). Increased urbanization, enabled by the harnessing of fossil energy for industrialization and the rise of capitalism, along with democracy, provided people with opportunities to interact in ways not possible in rural areas. These opportunities included becoming educated, developing civil society, creating new mechanisms for self-governance through voting and public institutions, and creating art and literature, fashion and places for convivial interaction such as restaurants and public parks. Cities created spaces for difference, for people who, previously, might have been ostracized, banished, burned (like 'witches' or heretics) or otherwise punished. They enabled the flourishing of the human spirit and to develop, human potential in ways not possible in the countryside. The city makes room for the moral levelling of difference, as the great sociologist Howard Becker (1991) points out in his discussion of deviance. He pointed out that often deviance, upon inspection and situation in the urban where it could exist beyond a small homogeneous group, was nothing more deviant than what others did. Cities provide an environment where art making, for example – music painting, literature, films – is a social enterprise in which a huge range of people play essential roles to produce an artefact that social groups then dignify or acknowledge as art. Cities allow collective actions that go beyond reproduction of daily life, as is often the case in rural communities. They allow many more different roles, they foster society and the growth of not only the individual but collective creation and innovation. Clearly, there is an important element of economic opportunity in cities and opportunities for different careers and pathways but a career or pathway may be defined by its creative output and its contribution to the community as well as its contribution to the economy.

The modernist city, so well described and dissected by Graham and Marvin (2001), co-evolved with new governmental systems predicated on democratic participation. The governmental apparatus grew alongside the technosphere, with dedicated agencies and departments, elected and appointed officials, and budgeting of revenues based on taxes – enlisting residents' wealth to create the infrastructure and management systems that served all equitably (in theory). That is, anywhere in a modern city, residents could access the same services such as clean water, safe buildings and sanitation services. All of these forms and processes were new, scientific and rational, aimed at making cities more functional, efficient and sanitary. The changes had, at least in the United States, the backing of the business sector, as they provided much needed infrastructure to facilitate commerce and manufacturing. But, the ability to harness energy and to distribute it throughout the urban fabric and to use it to extract building materials and to transform them, to transport them and to pump and push energy and water and move people and goods, was essential to the rapid expansion of this new kind of city. Figures of urban growth over a few decades are staggering. Resources from increasingly distant hinterlands were accessed, as William Cronon shows in his marvellous book, *Nature's Metropolis: Chicago and the Great West* (1991). And, of course, the environmental impacts of this increasing concentration of humanity, rising economic activity and affluence have cumulated in the contemporary period with climate change, water stress, pollution and other impacts.

The post carbon city

The post carbon city opens up the potential for rethinking and recapturing the role of cities beyond the economic workshops of the global economy to create places that nurture human well-being, creativity, self-governance, and purposeful work that is aimed at meeting human needs first, not the continual expansion of the gross domestic product. Looking forward, thinking about the post carbon city, what are some of the considerations to examine?

Infrastructure

Modernist infrastructure and large centralized systems dependent on fossil fuels have managed to erase space and geography. Many cities across the world employ the same types of infrastructures based on engineering science that has normalized standards and generally use centralized systems. For example, sewage is collected into centralized plants that then release treated water into receiving waters. Stormwater infrastructure is based on grading streets and parking lots to direct water into street drains to go either to sewage treatment plants or to discharge into receiving waters. City engineers apply these principles according to density, population numbers, soils and topography, the street networks and lot grid, and of course, funding.

Nevertheless, these are not systems engineered to optimize systems using local climate, soils, wind and sun, vegetation, or other biogeographical conditions. Cities in dry climates have not been engineered to capture stormwater or reuse grey or black water, for example. Cities in sunny climates do not generate electricity using the sun for even part of their need, although this is beginning to change. Modernist cities apply human engineered grey infrastructure in a spaceless geography where concrete, pipes, wires, technologies and fossil energy are deployed to resolve urban needs. This is a highly materials and energy intensive strategy.

Developing cities aspire to this model. Electricity and natural gas distribution networks operate the same across cities, as well as heating and air conditioning strategies. Cities in cold climates bring in heating oil or utilize district heating that is primarily dependent on the use of fossil energy. Cities in humid or arid climates use air conditioning and may also import water from long distances, as Los Angeles does. Building materials have become more homogeneous and available, transported far distances with little regard to their life span and reuse, nor their scarcity. There are exceptions: for example, aggregate to make concrete, in part because of its weight and bulkiness, tends to be mined nearer to its transformation into concrete, but wood products are imported from far-flung places.

With less carbon energy available, each city in each geographical region will have to truncate those long supply chains and re-evaluate the embedded energy in the urban fabric (Reyna and Chester 2015). Building materials and energy sources will change. Cities will need to be redesigned to contend with local environmental conditions and will begin to look increasingly different one from another. Building materials will be sourced more locally and those that are not will be expensive and rare. Existing building materials will be reused and there will be materials mining in solid waste dumps. The post carbon city is likely to become a zero waste city as the costs of new materials sourced from far off places will be exorbitantly expensive.

The post carbon city is a de-modern city, a city in which energy and material intensive grey infrastructure, gets supplemented, complemented and patched with other types of infrastructure, including distributed infrastructure that captures stormwater and reinfiltrates it, captures wastewater, treats it and puts it back in the water system and captures the 'waste' as food – fertilizer, biogas, and compost. De-modern connotes a state of being that recognizes the historically contingent notion of modern, one that puts in contrast places that are on the development path from those that are already modern and naturalizes that trajectory. Chakrabarty discusses modernity as a stagist theory that has become largely accepted and totalizing (2000). Thinking in *de-modern* terms is a way of breaking free of this historical legacy. Post carbon cities will no longer be able to overpower local conditions through the application of fossil energy and will need intelligence of place to redesign infrastructures to take advantage of local opportunities and resources – slopes, sun and wind, vegetation and water.

But such infrastructure presupposes a change in the governance of land uses, and a redistribution of buildings on the ground such that the urban fabric allows and enables the use of local resources. It will involve new urban forms aimed at reducing building energy use and maximizing the resource potentials of the local environment. In some areas vast amounts of nearby land may need to be devoted to forests for fuel and materials. In others, land, roofs and other surfaces will be devoted to producing electricity. A one-size-fits-all engineered infrastructure will no longer suffice. This new form could change the way the boundaries of cities are thought of beyond either city and countryside or city and suburb. Rather it would make for immediately relevant and important working landscapes and new experimental boundary drawing. In early cities, there was a clear demarcation between the city and the countryside, but cities and populations were small and many agriculturalists could walk to the surrounding fields. If cities in the future need to dramatically shorten supply chains, then what are the new political and management geographical jurisdictions? Perhaps it is an opportunity to make political geographical boundaries more in sync with 'natural' ones such as watersheds.

Food

Globalization and cheap fossil energy has revolutionized food production and distribution. Today, the northern hemisphere can be supplied by the southern hemisphere (rarely the other way around due to wealth distribution), and in the north, consumers can eat cherries (for example) in the winter. Tomatoes can be eaten year round and there has been an acceleration of concentration of size of farms and ownership of key choke points including seeds, fertilizers, animal feed and medicines, technical know-how, farm credit and distribution chains. This concentration has been supported by regulatory regimes like the European Union, free trade agreements, and national regulations such as those promulgated by the European Union and the US Food and Drug Administration and the US Department of Agriculture, setting standards that essentially require engagement with the globalized food system. But much of this requires easily available fossil energy, whether to manufacture fertilizer, drive the large tractors necessary to farm thousands of acres of land, homogenize milk and cheese, process grain, or frack the corn kernel (Pollan 2006), and so much more.

The concentration of ownership has been accompanied in many places with a rise in prices of agricultural land, and in areas near cities, additional land speculation has added to the cost of land. But in a post carbon world, shipping cherries from Chile to Washington, London or Paris in the winter will be more difficult and most likely far more costly. Tropically raised products such as coffee will be rare and expensive. Older patterns of growing food closer to markets will have to return and techniques of food storage will change too. Pickling, canning, salting, drying, as well as limited freezing (depending on energy availability), will be necessary but mostly there will need to be affordable agricultural land in the nearby areas, thus potentially requiring land reform. Land's highest and best use may no longer be speculative gain and it will need to be priced and managed accordingly. Food habits will change and if there is not enough room to grow and feed livestock, then meat eating will decline.

The agricultural supply chain is a critical one for the post carbon city and raises questions about the ability of land resources to feed mega cities without fossil energy and long import supply chains. It would be important to study, for example, existing food supply chains to megacities in the southern hemisphere (e.g. Mexico City or Lagos). Is there currently food growing and distribution in those cities? How much food is imported, from where and what is the fossil fuel content? How much hunger is there? What is the water availability for local agriculture, and its safety, is land contaminated? And is hunger a matter of absolute food scarcity or simply poverty?

Land speculation in the west and the growth of cities have driven farmers farther and farther away from cities, forcing them to become increasingly dependent on supply chains they do not control. The connection to the city is ruptured, except for the emerging farmers' market phenomenon that few access relative to population size and where the food is often expensive. In the post carbon city, it is likely food will be more expensive as it will be more labour intensive and support nearby farmers. How will that be counterbalanced so urbanites can afford to eat? There are deep questions of social organization once it is ascertained that there can be enough food provided in a relatively regional geography.

Economy

The economic activities of post carbon cities will need to be reorganized to provide for people's local needs; productive activities and trade will be based on local biophysical assets and knowledge, along with fuel constraints (wind, water, and energy storage). Patterns of trade and its pace will need to evolve accordingly, like for food products. Specialization will no doubt both change and re-emerge depending on local resources and skills. There will need to be a wider diversity of skills to provide for human needs: skills like growing food, building from more locally sourced materials, paper making, manufacturing with local energy and materials, or recycled materials. Reskilling will be an important part of the economy and education. At the same time, the scale and concentration of ownership will no longer be sustained and new, different, and also old forms will develop. Diversity of ownership and enterprise types can emerge as a fossil fuel based economy contracts. These will need to be created from the ground up, transitioning from – in the west – relying on imported consumer items to more local and regional manufacturing and production. Common Weal (2015), a Scottish organization, describes the new economy in this way:

> A Common Weal economy would begin by getting local trade right. Local businesses will be much better at supplying each other and will take the lead in supplying public bodies with local produce with much more emphasis placed on the need to keep business local. This will lead to an expansion of regional food production (rather than speculative housing developments), for example, with many more growers supplying local bakers, microbreweries, artisan food makers and so on, all selling to local consumers.

Commonweal goes on to propose that less imported consumer goods and more manufacturing at many different scales: small, specialist manufacture such as micro-engineering and prototyping using non-fossil-fuel power sources. 'Smart specialisation' will develop where specializing into high-skill areas can be traded in a post carbon fuel economy, such as developing hydrogen-powered electric ships. A post carbon economy will be built on the careful (re)utilization of resources – natural, human and geo-strategic: land, wind, tide, wave, water – and skill, innovation, creativity and design skills (Common Weal 2015).

In this economy, education becomes critical, and a significant purpose of the post carbon city would be to advance knowledge, increase educational opportunities, and link residents to creative and fulfilling work and activities. In some ways, the economy becomes a facet of civil society, re-embedded in society, post capitalist as it is post carbon.

Land and property

Any one of the changes described above will require a fundamental change in property rights, a revival of the commons and new forms of decision-making. For cities to adjust to no longer being dependent on far-flung resources harnessed through the deployment of fossil energy, more local and regional resources will be necessary, including, for example, rooftops for solar power, or urban land that is now private for water infiltration. Currently, this depends on property-owner by property-owner devoting their land – or rooftop – to these uses. A post carbon city will require access to these resources for energy production and water reinfiltration; owners will not be able to choose whether to participate. Private lots may need to be commandeered to install battery storage units or other power storage technologies.

Other deep changes are possible. For example, with the effects of climate change already underway, such as sea level rise, private property will no longer be allowed in vulnerable coastal locations. But currently, the scale and scope of this type of private property retirement, currently subject to compensation by government in the United States under the Constitution, overwhelms any fiscal ability to do so. The rights to compensation for the public interest – restoring coastal dunes for example to protect urban areas in land – will have to be altered such that the common good is served and cities can adapt to the impacts of the legacy of burning fossil fuels. This will, in the United States at least, strike at the very heart of the US Constitution as the Fifth Amendment protects private property rights from takings by the government without just compensation and due process. This protection, while ensuring property owners that their land will not be arbitrarily seized, also stands in the way of adaptation to climate change and transformation of the urban fabric for a new post carbon infrastructure and urban form. In order to establish non-arbitrary property seizures, new governance organizations and methods of compensation will need to evolve.

Current property regimes of homeowners associations that, for example, prohibit clothes drying out of doors, require lawns, prescribe the colour of buildings, and generally codify practices that are energy intensive, will be replaced by ones that are aimed at energy savings. These could include the requirement that all buildings must be painted white to reflect sunlight in warm climates or be built to capture solar gain in the winter. The changes needed are multiple, but the challenge will be governance. Thus the concept of the commons – beyond mere property rights enforcement – will have to be revisited and reinvented for a post carbon city. The transformations necessary for the post carbon city reintroduce the land question and land reform probabilities.

Governance

Imagining post carbon urban governance is challenging, as urban governance is nested and tiered in other layers of government and is constrained by international, national, state (or province) laws, codes and procedures. Cities are not autonomous spheres, although they have been at the forefront of addressing climate impacts (Bulkeley *et al.* 2013). Socio-technical challenges to transition away from carbon intensive systems have been examined – listing and describing all of the actors and agencies that will have to be engaged with this shift (Geels 2004; Hodson and Marvin 2009; Bulkeley *et al.* 2013). But the overall governance system is more elusive.

In part, this is because the problem is not posed as a shift away from cities as economic engines of global capitalism that have co-evolved economic growth with governance systems. Governance systems today are organized around private property protection and creating good conditions for economic activity, including reliable infrastructure. Infrastructure transition analysis is, in some ways, concerned with maintaining these same reliable systems – uninterrupted electricity, but produced cleanly, for example. Struggle between interests about future resource requirements, infrastructures, and social interests is recognized, but over what goals? Again, there is acknowledgement of the important role of cities in transition approaches (Hodson and Marvin 2011) but what cities do and what cities are for is not part of the exploration. There is acknowledgement that consumption needs to be curbed (Dhakal 2013) but how this is reconciled with capitalism's need for growth and the role of cities in employing people in producing the very goods that are consumed still needs greater attention. Thus, there is need for attention to what new forms of governance are needed to usher in post carbon cities that have goals that diverge from maintaining current systems, albeit powered with different sources of energy. Questions that need to be addressed include: will a post carbon future engender the disintegration of current nation states and the rise anew of city-states that source their needs from their own hinterlands, engaging in sophisticated, but limited exchange with other city states across the globe? Indeed, is post carbon a pathway that demodernizes the city and its current organizational structure, both physical and governmental?

Conclusions

Clearly, many of the changes that will be needed require national and international rule changes, from the property protection clauses of the US Constitution to the General Agreement on Trade and Tariffs. Relationships amongst local, regional, state, nation and international levels will be highly upset with the decline of the fossil fuelled global economy. Systems will unravel and be replaced with different ones, but, most importantly, the power of fossil energy as a singular force that has enabled the transformation of the planet will most likely not be replaced. Centralized systems and infrastructures have been enabled by fossil energy. This is not to say that large infrastructures have never existed before – the Great Wall of China and Roman aqueducts as well as large-scale empires have all harnessed resources and people. But previous to fossil fuels, these were built on human labour (often slave labour). A post carbon city that functions without such labour is a new form. Arriving there will necessitate rethinking the role of cities in the capitalist era, an era that will no longer be modern.

It is tempting to turn to the question of transition, or whether there are any ongoing experiments that are leading toward a new type of city, one that is post carbon. One might turn to some of the reappropriations of urban spaces in declining cities, such as Bologna, Detroit or even parts of Tokyo, to try to perceive what a post carbon city might look and feel like. In these places, cultural experiments are taking place: squatting of buildings and creating of commons and communes, the construction of community gardens, 'maker' spaces where people can cook, process food, sew, fix bicycles and do woodworking together, and with an emphasis on the role of art and artistic creation. These are not capitalist enterprises but rather communal shared enterprises where time takes on a different dimension and the goal is not efficiency but rather sociability, quality, craftsmanship and learning. Sharing is also valued.

Whether these presage the civil society of a post carbon city is questionable, but voluntary movements in cities in decline that seek to create quality of life are an interesting development and certainly not predicated on carbon or the circulation of capital. What Occupy Wall Street and these movements of cultural opposition point to is the need to attend to a deeper set of questions about our social, political and material existence if a wider change is to be achieved. It includes discovering or rediscovering values such as compassion, sympathy and joy. It includes expanding space and time for creativity and sociability and the satisfaction of rewarding and fulfilling work. Re-embedding the economy in society will be an important aspect of moving toward a post carbon city. Expanding our conception of city and acknowledging its potential beyond an economic workshop is a first and critical step toward decarbonization.

References

Becker, H. (1991). *Outsiders: Studies in the Sociology of Deviance*. New York: Free Press.

Brundtland, G.H. (1987). *Our Common Future, Report of the World Commission on Environment and Development*. Geneva: United Nations.

Bulkeley, H., Castán Broto, V., Hodson, M. and Marvin, S. (eds) (2013). *Cities and Low Carbon Transitions*. New York: Routledge.

Callenbach, E. (1975). *Ecotopia*. Berkeley, CA: Banyan Books.

Chakrabarty, D. (2000). *Provincializing Europe*. Princeton, NJ: Princeton University Press.

Common Weal (2015). *Common Weal website*. [Online]. Available: http://www.allofusfirst.org [25 September 2015].

Cronon, W. (1991). *Nature's Metropolis, Chicago and the Great West*. New York: W.W. Norton.

Dhakal S. (2013). Urban energy transitions in Chinese cities. In Bulkeley, H., Castán Broto, V., Hodson, M. and Marvin, S. (eds), *Cities and Low Carbon Transitions*. New York: Routledge, 73–87.

Draper, H. (1978). *Karl Marx's Theory of Revolution*. New York: Monthly Review Press.

Folke, C., Kautsky, N., Berg, H., Jansson, A. and Troell, M. (1998). The ecological footprint concept for sustainable seafood production: a review. *Ecological Applications*, 8: S63–S71.

Garrod, G. and Willis, K.G. (1999). *Economic Valuation of the Environment: Methods and Case Studies*. Cheltenham: Edward Elgar.

Geels, F.W. (2004). From sectoral systems of innovation to socio-technical systems: insights about dynamics and change from sociology and institutional theory. *Research Policy*, 33: 897–920.

Graham S. and Marvin S. (2001). *Splintering Urbanism: Networked Infrastructures, Technological Mobilities and the Urban Condition*. New York: Routledge.

Hodson, M. and Marvin, S. (2009). Cities mediating technological transitions: understanding visions, intermediation and consequences. *Technology Analysis and Strategic Management*, 21: 515–534.

Hodson M. and Marvin S. (2011). Can cities shape socio-technical transitions and how would we know if they were? In Bulkeley, H., Castán Broto V., Hodson M. and Marvin S. (eds), *Cities and Low Carbon Transitions*. New York: Routledge, 42–53.

Ibanez, D. and Katsikis, N. (2014). Editorial: grounding metabolism. *New Geographies*, 6: 2–9.

Kallis, G., Demaria, F. and D'Alisa, G. (2015). Introduction: degrowth. In D'Alisa, G., Demaria, F. and Kallis. G. (eds), *Degrowth: a Vocabulary for a New Era*. New York: Routledge, 1–18.

Kates, R.W. (2010). *Readings in Sustainability Science and Technology*. CID Working Paper no. 213. Cambridge, MA: Center for International Development at Harvard University.

Kaufmann, R. and Cleveland, C. (2007). *Environmental Science*. New York: McGraw-Hill Science.

Klein, N. (2014). *This Changes Everything*. New York: Simon and Schuster.

Kong, F., Yin Waiwei, M. and Nakagoshi, N. (2007). Using GIS and landscape metrics in the hedonic price modeling of the amenity value of urban green space: a case study in Jinan City, China. *Landscape and Urban Planning*, 79: 240–252.

Lerner, J. (2014). *Urban Acupuncture*. Washington DC: Island Press.

McGranahan, G. and Satterthwaite, D. (2003). Urban centers: an assessment of sustainability. *Annual Review of Environmental Resources*, 28: 243–74.

McNeill, J.R. (2000). *Something New Under the Sun: An Environmental History of the Twentieth-Century World*. New York: W.W. Norton.

Mol, A.P.J., Sonnenfeld D.A. and Spaargaren G. (eds) (2009). *The Ecological Modernisation Reader*. New York: Routledge.

National Research Council (2013). *Urban Forestry Toward an Ecosystem Services Research Agenda, A Workshop Summary*. Washington, DC: The National Academies Press. Pataki, D.E., Carreiro, M.M., Cherrier, J., Grulke, N.E., Jennings, V., Pincetl, S. and Zipperer, W.C. (2011). Coupling biogeochemical cycles in urban environments: ecosystem services, green solutions, and misconceptions. *Frontiers in Ecology and the Environment*, 9: 27–36.

Pincetl, S. (2012). Nature, urban development and sustainability – what new elements are needed for a more comprehensive understanding? *Cities*, 29: S32–S37.

Pincetl, S., Chester, M., Circella, G., Fraser, A., Mini, C., Murphy, S., Reyna, J. and

Sivaraman, D. (2014). Enabling future sustainability transitions: an urban metabolism approach to Los Angeles. *Journal of Industrial Ecology*, 18: 871–882.

Pollan, M. (2006). *The Omnivore's Dilemma: A Natural History of Four Meals*. New York: Penguin Press.

Portney, K.E. (2003). *Taking Sustainable Cities Seriously: Economic Development, the Environment, and Quality of Life in American Cities*. Cambridge, MA: The MIT Press.

Register, R. (2002). *Ecocities, Building Cities in Balance with Nature*. Berkeley, CA: Berkeley Hills Books.

Reyna, J. and Chester, M. (2015). The growth of urban building infrastructure and its unintended lock-in and embedded environmental effects, *Journal of Industrial Ecology*, 19: 524–537.

Ritthoff, M., Rohn, H. and Liedtke, C. (2002). *Calculating MIPS: Resource Productivity of Products and Services*. [Online]. Available: http://wupperinst.org/en/publications/details/wi/a/s/ad/584/ [23 April 2014].

Smil, V. (2008). *Energy in Nature and Society: General Energetics of Complex Systems*. Cambridge, MA: The MIT Press.

Storper, M. (2013). *Keys to the City: How Economics Institutions, Social Interactions and Politics Shape Development*. Princeton, NJ: Princeton University Press.

Wackernagel, M. and Rees, W. (1996). *Our Ecological Footprint: Reducing Human Impact on the Earth*. Philadelphia, PA: New Society Publishers.

Williams, J.H., Haley, B., Kahrl, F., Moore, J., Jones, A.D., Torn, M.S. and McJeon, H. (2014). *Pathways to Deep Decarbonization in the United States. The US Report of the Deep Decarbonization Pathways Project of the Sustainable Development Solutions Network and the Institute for Sustainable Development and International Relations*. [Electronic]. Available at: https://ethree.com/publicaitons/index US2050.php [23 April 2014].

Index

Note: page numbers in **bold** indicate figures or tables.